T0226304

Lecture Notes in Computer Science

Lecture Notes in Computer Science

Edited by G. Goos and J. Hartmanis

121

Zahari Zlatev
Jerzy Wasniewski
Kjeld Schaumburg

Y12M

Solution of Large and Sparse Systems of
Linear Algebraic Equations
Documentation of Subroutines

Springer-Verlag
Berlin Heidelberg New York 1981

Authors

Zahari Zlatev
Computer Science Department, Mathematical Institute
University of Aarhus, Ny Munkegade, DK 8000 Aarhus C

Jerzy Wasniewski
The Regional Computing Centre at the University of Copenhagen
Vermundsgade 5, DK 2100 Copenhagen

Kjeld Schaumburg
Department of Chemical Physics, University of Copenhagen
The H.C. Oersted Institute, Universitetsparken 5,
DK 2100 Copenhagen

AMS Subject Classifications (1979): 15 A 04, 15 A 06, 15 A 23, 65-04,
65 F 05, 65 F 10, 68 E 05
CR Subject Classifications (1981): 5.14, 4.6

ISBN 3-540-10874-2 Springer-Verlag Berlin Heidelberg New York
ISBN 0-387-10874-2 Springer-Verlag New York Heidelberg Berlin

CIP-Kurztitelaufnahme der Deutschen Bibliothek
Zlatev, Zahari:
Y12M [YM] solution of large and sparse systems of linear algebraic equations:
documentation of subroutines / Zahari Zlatev; Jerzy Wasniewski; Kjeld Schaumburg.
– Berlin; Heidelberg; New York: Springer, 1981.
(Lecture notes in computer science ; 121)
ISBN 3-540-10874-2 (Berlin, Heidelberg, New York);
ISBN 0-387-10874-2 (New York, Heidelberg, Berlin)
NE: Wasniewski, Jerzy:; Schaumburg, Kjeld:; GT

Printing and binding: Beltz Offsetdruck, Hemsbach/Bergstr.
2145/3140-543210

Preface

The Y12M is a package of Fortran subroutines for the solution of large and sparse systems of linear algebraic equations developed at the Regional Computing Centre at the University of Copenhagen (RECKU). Gaussian elimination and pivotal interchanges are used to factorize the matrix of the system into two triangular matrices L and U. An attempt to control the magnitude of the non-zero elements in order to avoid overflows or underflows and to detect singularities is carried out during the process of factorization. Iterative refinement of the first solution may be performed. It is verified (by a large set of numerical examples) that iterative refinement combined with a large drop-tolerance and a large stability factor is often very successful when the matrix of the system is sparse. Not only is the accuracy improved but the factorization time is also considerably reduced so that the total computing time for the solution of the system with iterative refinement is less than that without iterative refinement (in some examples the total computing time was reduced by more than three times). The storage needed can often be reduced also.

Note that if the matrix of the system is dense then the total computing time for the iterative solution of the system is always larger (because extra time must be used to perform the iterations needed to improve the accuracy of the first solution). Note too that the use of iterative refinement with dense matrices leads to an increase of the storage by a factor approximately equal to 2 (because a copy of the matrix of the system must be kept).

The factorization found by the use of large values of the drop-tolerance is sometimes referred to as incomplete. The matrix LU obtained in this way can be considered as a preconditioned matrix. Preconditioned matrices are often used when systems with symmetric and positive definite matrices are solved by the conjugate gradients method. When the matrices are general the conjugate gradients method can not be used (at least in its classical form). In our package iterative refinement is used instead of the conjugate gradients method. The experiments show that this approach is normally extremely efficient. This is especially true for expensive problems, i.e. problems whose matrices are such that many fill-ins are produced in the process of the LU decomposition. It should be mentioned that efficency is most required just for such problems.

It is necessary to emphasize that a reliable error estimation will normally be obtained when the iterative refinement process is convergent. No error estimation can be found when the system is solved directly. However, one can expect that the required accuracy will be achieved, especially when double precision is used, if the condition number of the coefficient matrix is not extremely large. A subroutine which evaluates the condition number of a matrix is available and may optionally be called. If this subroutine is called (which can be done when the LU decomposition is calculated), then a reliable measure of the sensitivity of the results to the round-off errors will be obtained. It must be emphasized that the use of the subroutine for evaluation of the condition number is relatively cheap; its computational cost is equal to the computational cost for two back substitutions.

There exist problems for which the application of iterative refinement is not very efficient with regard to the storage and computing time used (e.g. when many systems with the same coefficient matrix are to be solved). Therefore the iterative refinement process should in our opinion be only an option in the package for the solution of large and sparse systems of linear algebraic equations. Using some machine dependent facilities (paging, multibanking etc.) one can modify the iterative refinement option so that it will never use more storage than the direct solution option. The price which should be paid for this is a modest increase of the computing time. A modified version of the iterative refinement option which uses multibanking has been developed for Univac 1100 series computers at RECKU (the Regional Computing Centre at the University of Copenhagen).

All subroutines of the package have been run on three different computers: a Univac 1100/82 computer at the Regional Computing Centre at the University of Copenhagen

(RECKU), an IBM 3033 computer at the Northern Europe University Computing Centre (NEUCC) and a CDC Cyber 173 computer at the Regional Computing Centre at the University of Aarhus (RECAU). Some results from these runs are reported in this book.

If reference is made to a paper or a book on some page, then its title, the name(s) of the author(s) and the place of publication appear as a footnote on the same page. Thus the superscript in any reference indicates the number of the footnote where a detailed information about the reference is given. All references are also listed at the end of this book. The INDEX can be used to check the page where a reference to any paper or book is made.

The codes of the subroutines with full documentation are in RECKU Library. They are available at the usual costs (for the magnetic tape, machine time, shipment, etc.). The requests should be addressed to J. Wasniewski. Advances in theory and the experience of users may prompt alterations in these codes. Readers and/or users are invited to write to the authors concerning any changes they may advocate.

CONTENTS

1. Introduction to Y12M

1.1. Scope of the Y12M

Y12M is concerned with the calculation of the solution of systems of linear algebraic equations whose matrices are large and sparse.

Two matrices A and B are of the same structure
if $a_{ij} \neq 0$ implies $b_{ij} \neq 0$ and $b_{ij} \neq 0$ implies $a_{ij} \neq 0$.

The following notation is useful in the classification: matrices denoted by the same letters and subscripts are the same, matrices denoted by the same letters and different subscripts are of the same structure (but different), matrices denoted by different letters are of different structure.

We consider the solution of the following problems:

(i)   Only one system with a single right-hand side, $Ax = b$, is to be solved.

(ii)   A sequence of systems with the same matrices is to be solved. The matrices of this sequence are

$$A_1, A_1, ..., A_1.$$

The case where one system with many right-hand sides is to be solved can easily be presented as a sequence of systems with the same matrices.

(iii)   A sequence of systems whose matrices are different but of the same
          structure is to be solved. This means that the matrices of this sequence are

$$A_1, A_2, ..., A_p.$$

(iv)   A sequence of systems whose matrices are of the same structure and
          some of them (but not all of them) are different is to be solved. This means
          that the matrices of this sequence are

$$A_1, A_1, ..., A_1; A_2, A_2, ..., A_2; ...; A_p, A_p, ..., A_p.$$

(v)   A sequence of systems whose matrices are of a different structure is to be
          solved. The matrices of this sequence are

$$A, B, C, ..., Z.$$

There are recommendations for the use of the package in each of the above five cases.

1.2. Background to the Problem

1.2.1. Storage Operations

Given a matrix A with Z non-zero elements. These elements are stored (ordered by rows) in
the first Z positions of array A. Their column numbers are stored in the same positions of
array SNR. Arrays A and SNR form the row ordered list. The row numbers of the non-zero
elements (ordered by columns) are stored in the first Z positions of array RNR. Array RNR

forms the column ordered list. This storage algorithm was first proposed by Gustavson[1][2] . During the computation the non-zero elements that are not necessary in subsequent stages are removed from the lists. Unfortunately, some new non-zero elements (fill-ins) are created and placed at the beginning or at the end of the row (column) if there are free locations. Otherwise, a new copy of the row (column) at the end of the row (column) ordered list is made, thus freeing the locations originally reserved for the row (column). Obviously there is a limit to the number of new copies that can be made without exceeding the capacity of the arrays, and therefore occasional "garbage" collections are necessary. More details about the storage scheme used in our subroutines can be found in Zlatev, Schaumburg and Wasniewski[3] and Zlatev[4] .

1.2.2. Mathematical Method

Gaussian elimination is used with interchanges. The system

$$PAQ(Q^Tx) = Pb$$

(where P and Q are permutation matrices) is replaced by

$$LU(Q^Tx) = Pb$$

1.   Gustavson, F.G. - "Some Basic Techniques for Solving Sparse
Systems of Linear Equations",
In: "Sparse Matrices and Their Applications" (D.J. Rose and R.A. Willoughby, eds.),
pp. 41-52, Plenum Press, New York, 1972.

2.   Gustavson, F.G. - "Two Fast Algorithms for Sparse Matrices:
Multiplication and Permuted Transposition",
ACM Trans. Math. Software, 4, pp. 250-269, 1978.

3.   Zlatev, Z., Schaumburg, K. and Wasniewski, J. - "Implementation
of an Iterative Refinement Option in a Code for Large and Sparse Systems".
Computers and Chemistry, 4, pp. 87-99, 1980.

4.   Zlatev, Z. - "Use of Iterative Refinement in the Solution
of Sparse Linear Systems", Report 1/79, Institute of Mathematics
and Statistics, The Royal Veterinary and Agricultural University,
Copenhagen, Denmark, 1979 (to appear in SIAM J. Numer. Anal.).

(where L is a unit lower triangular matrix and U is an upper triangular matrix). Then the latter system is solved by forward and back substitution and (normally) an approximation to the solution vector x is computed. If iterative refinement is to be used then we denote the above solution vector by x_1 and perform the following successive calculations:

$$r_{k-1} = b-Ax_{k-1},$$

$$d_{k-1} = QU^{-1}L^{-1}Pr_{k-1},$$

$$x_k = x_{k-1}+d_{k-1},$$

where k = 2, 3, 4, ...

(various stopping criteria must be used in order to terminate the process if the required accuracy is achieved or if the process is not convergent). Normally the accuracy will be improved if iterative refinement is applied in the calculations of the solution of linear systems.

More details about the theory of the Gaussian elimination can be found e.g. in Forsythe and Moler[5] , Stewart[6] and Wilkinson[7][8] .

5. Forsythe, G.E. and Moler, C.B. -
"Computer Solution of Linear Algebraic Equations",
Prentice-Hall, Englewood Cliffs, N.J., 1967.

6. Stewart, G.W. - "Introduction to Matrix Computations",
Academic Press, New York, 1973.

7. Wilkinson, J.H. - "Rounding Errors in Algebraic Processes",
Prentice-Hall, Englewood Cliffs, N.J., 1963.

8. Wilkinson, J.H. - "The Algebraic Eigenvalue Problem",
Oxford University Press, London, 1965.

1.3. Recommendations on the Use of the Routines

1.3.1. Choice of Pivotal Strategy

Interchanges are normally used in order to preserve the sparsity of the original matrix (i.e. to minimize the number of non-zero elements created during the factorization) and to ensure numerically stable computations (i.e. to attempt to prevent the occurrence of large errors in the solution vector).

The choice of the pivotal strategy depends on matrix A. We have three pivotal strategies, each of which may be used by appropriate initialization of the parameter IFLAG(3) before the call of the package. For general matrices A, IFLAG(3) = 1 must be used. In this case the number of rows that will be investigated at each stage of the elimination in order to determine the pivotal element must be given as well. This number must be initialized in IFLAG(2) and should not exceed three. A generalized Markowitz strategy is used[9] , i.e. among the elements of the selected IFLAG(2) rows with least numbers of non-zero elements, the element which satisfies the stability requirement and for which the product of the other non-zero elements in its row and the other non-zero elements in its column is a minimum will be chosen as a pivotal element. Moreover, if there are several such elements, the largest in absolute value will be chosen (more details can be found in Zlatev[9] and Zlatev, Schaumburg and Wasniewski[10]).

If the use of pivotal elements on the main diagonal is sufficient (and this is the case if e.g. the matrix is symmetric and definite or diagonally dominant) then the package can take advantage of this property. IFLAG(3) should be initialized with the value 2 before the call of

9.   Zlatev, Z. - "On Some Pivotal Strategies in Gaussian Elimination
        by Sparse Technique", SIAM J. Numer. Anal., 17, pp. 18-30, 1980.

10.   Zlatev, Z., Schaumburg, K. and Wasniewski, J. - "Implementation
        of an Iterative Refinement Option in a Code for Large and Sparse Systems".
        Computers and Chemistry, 4, pp. 87-99, 1980.

Y12M.

In rare cases no pivoting is necessary. The sparsity pattern of the matrix and the numerical stability of the computation will be preserved if no pivotal interchanges are made. In this case IFLAG(3) should be initialized with the value zero before the call of the Y12M. This choice applies when matrix A is symmetric and definite or diagonally dominant and moreover, nearly all non-zero elements are located not far from the main diagonal. When the iterative refinement (IR) option is used then computations without pivoting are possible for quite general matrices. This has successfully been demonstrated by Schaumburg et al.[11] .

Note that the use of the two special strategies may reduce the computing time considerably. In our experiments with positive definite matrices and with the use of the above special strategies, the computing time needed to solve the systems by our package is comparable to the computing time needed to solve the same systems with codes especially written to be used with positive definite matrices, see e.g. Zlatev, Wasniewski and Schaumburg[12] .

1.3.2. Robustness

In our opinion the code for the solution of large and sparse systems of linear equations should attempt to detect any of the following situations: (a) the elements of the matrix grow too quickly (then the subsequent computations are not justified, and even overflow may take place), (b) very small elements appear in the computation (so that underflows are possible), (c) the matrix is singular (or nearly singular, so that the machine accuracy will not be sufficient to compute an acceptable solution).

11. Schaumburg, K., Wasniewski, J. and Zlatev, Z. - "The Use of Sparse Matrix Technique in the Numerical Integration of Stiff Systems of Linear Ordinary Differential Equations". Computers and Chemistry, 4, pp. 1-12, 1980.

12. Zlatev, Z., Wasniewski, J. and Schaumburg, K. - "Comparison of Two Algorithms for Solving Large Linear Systems". Report No 80/9, Regional Computing Centre at the University of Copenhagen, Vermundsgade 5, DK-2100 Copenhagen, Denmark, 1980.

Below we describe how these problems are handled by the routines.

(a). Denote $a = \max|a_{ij}|$ where a_{ij}, $i,j = 1,2,...N$, are the elements of the original matrix A. Denote $b_k = \max|a_{ij}^{(s)}|$ where $k = 1,2,....,$ N-1; $s = 1,2,....,k$; $i,j = s,s+1,...,N$ and $a_{ij}^{(s)}$ are elements of matrix A which are to be transformed by the Gaussian transformations at stage s of the elimination. Let u be the stability factor (where $u \geq 1$ and u should be initialized in AFLAG(1) before the call of the package). It can then be shown that $b_k/a \leq (1+u)^k$ (for $k = 1,2,...,N-1$) and even examples where $b_k/a = (1+u)^k$ may be constructed. Moreover, during the computation we compute $\bar{L}\bar{U} = PAQ+E$ (instead of $LU = PAQ$) and for the elements e_{ij} of the pertubation matrix E it is true that $|e_{ij}| \leq 3.01b_{N-1}\varepsilon$ (where ε is the machine accuracy)[13] . Therefore it is clear that if b_k (stored by the subroutine Y12MC in AFLAG(7)) grows very quickly then the computed $\bar{L}\bar{U}$ will be inaccurate and in an extreme case overflow may occur. We monitor the largest computed elements and the growth factor b_k/a and if this factor is larger than a number prescribed by the user (this number should be initialized in AFLAG(3), we recommend 10^{16}) the package will stop the computation and give the error indication that the growth factor is too large. Note that smaller values of the stability factor AFLAG(1) tend to decrease the growth factor.

(b). Sometimes very small elements appear during the computation. The drop-tolerance T (stored in AFLAG(2)) may successfully be used in this situation. If during the computation $|a_{ij}^{(k)}| < T$ then the subroutine Y12MC will remove this element from the lists. In our experiments even the use of very small values of the drop-tolerance (say $T = 10^{-30}$) has proven very efficient in the cases where underflows were registered by the use of $T = 0$. Note that some compilers stop the computation and give error diagnostics if underflows appear. We recommend the use of $T = 10^{-12}$. It must be emphasized here that larger values for the drop-tolerance may be used with iterative refinement.

(c). Let the rank of matrix A be $k < N$. Then all elements in submatrix A_{k+1}, obtained after k stages of Gaussian elimination, should be equal to zero, but in general they are not because

13.   Reid, J.K. - "A Note on the Stability of Gaussian Elimination",
         J. Inst. Math. Appl., 8, pp. 374-375, 1971.

of round-off errors (the elements of matrix A_{k+1} are all elements whose row and column numbers are larger than or equal to $k+1$). Nevertheless, all the elements of A_{k+1} and the successive submatrices are normally small and it is possible (and in our examples it was so) that they will be removed and a row and/or a column without non-zero elements will be found (in this case IFAIL = 7 or 8 on exit will normally indicate that the matrix is numerically singular, except in the cases where the drop-tolerance is too large).

Another means of detecting singularity is the check for the minimal pivotal element (stored on exit in AFLAG(8)). One might check the whole sequence of pivotal elements (stored in array PIVOT, renamed Y in subroutine Y12MF). A very small pivotal element will indicate that the matrix is nearly singular (if AFLAG(1) is not too large). Note that if the absolute value of the current pivotal element is smaller than AFLAG(4) $\star$ AFLAG(6) then the routine stops the computation. A very small number must be initialized in AFLAG(4) before the entry (we recommend AFLAG(4) = 10^{-12}; the largest in absolute value element of the original matrix is stored by subroutine Y12MB in AFLAG(6)).

1.3.3. Storage of Matrix L

The principle of storing the non-zero elements of the unit lower triangular matrix L is automatically taken from the solution of systems with dense matrices (where this operation does not require extra storage and extra time). In the case of systems with sparse matrices the storage of L requires about 40% larger arrays A and SNR. Extra computing time is often needed too. Therefore L must be stored only if the problem justifies it (i.e. we have problems of type (ii) or (iv)) [14]. Moreover, note that some very large problems of type (i), (iii) or (v) may be solved without the use of secondary storage (as for example discs) only if the non-zero elements of matrix L are removed. If the non-zero elements of L are not required then IFLAG(5) = 1 must be initialized (the length NN of arrays A and SNR may be reduced in this case and NN = $2\star$Z will often be enough). If the non-zero elements are to be stored then

14.   See Section 1.1 "Scope of Y12M"

IFLAG(5) = 2 must be initialized before entry. If the non-zero elements of L are to be used (i.e. they were stored during a previous call of the Y12MC) then IFLAG(5) = 3 must be initialized before the call of the subroutines.

1.3.4. Matrices with the Same Structure

When a sequence of systems with the same structure is to be solved then the information about the pivotal elements used during the solution of the first matrix and the information about the non-zero elements of LU for the first system may be used in the solution of the next systems in the sequence (note that this may cause numerical instability if the non-zero elements of two successive matrices differ too much). During the factorization of the following systems no pivotal search is needed and, normally, no row and column collections and no moving of the non-zero elements of the row (column) to the end of the row (column) ordered list will be performed. The use of the old pivotal sequence of interchanges is always possible. However, the reordering of the non-zero elements, so that no collections and no movings of rows to the end of the row ordered list are needed, is only possible if the length NN of arrays A and SNR is larger than the number IFLAG(10). Similarly, the reordering of the row numbers of the non-zero elements (by columns), so that no collections and no movings of columns to the end of the column ordered list are needed, is only possible if the length of array RNR is larger than the number IFLAG(9). Both numbers, IFLAG(9) and IFLAG(10), have been computed by the routines during the solution of the first system in the sequence. Therefore we recommend the use of larger values of NN and NN1 if a complete use of the structure[15] is required. Note that if NN and/or NN1 are not larger than IFLAG(10) and/or IFLAG(9) the subroutines will normally solve all the systems (if the first one is solved). However, only a partial use of the structure will be made. See more details about the solution of systems whose matrices have the same structure in Zlatev, Schaumburg and Wasniewski[15].

15.   Zlatev, Z., Schaumburg, K. and Wasniewski, J. - "Implementation
       of an Iterative Refinement Option in a Code for Large and Sparse Systems".
       Computers and Chemistry, 4, pp. 87-99, 1980.

If a sequence of systems whose matrices are of the same structure is to be solved and if we want to use the information found during the solution of the first system then:

a) IFLAG(4) = 1 must be initialized before the call of the package for the first system of the sequence.

b) IFLAG(4) = 2 must be initialized before the call of the package for each of the next systems.

If occasionally numerical instability is detected (e.g. when the growth factor is very large) a new call of the package for the solution of the same system with IFLAG(4) = 1 should be performed. After that call, IFLAG(4) = 2 may be used again.

If systems whose matrices are not of the same structure (or are of the same structure but this property will not be exploited, because e.g. the non-zero elements are very different) then IFLAG(4) = 0 is the best choice. IFLAG(4) = 1 may be used with the call of each system but this will be sligthly more expensive.

1.3.5. Iterative Refinement

When systems whose matrices are dense are to be solved, the use of iterative refinement will normally improve the accuracy but more storage will be required (because a copy of the original matrix is preserved) and more computing time (because extra time is used to perform the iterations). When systems whose matrices are sparse are to be solved the use of iterative refinement will still normally improve the accuracy of the solution. Moreover, it is possible to introduce a relatively large drop-tolerance and a large stability factor[16] . The use of a large drop-tolerance tends to reduce both the storage needed and the factorization time (because

16.   See Section 1.3.2. Robustness.

all elements $a_{ij}^{(k)}$ which are smaller (in absolute value) than the drop-tolerance will be removed). Any element $a_{ij}^{(k)}$ in the selected rows with least numbers of non-zero elements may be chosen as a pivotal element at stage k of the elimination if the ratio of the largest in absolute value element in its row and the absolute value of the considered element $a_{ij}^{(k)}$ is smaller than the stability factor. Therefore it is clear that the use of large values for the stability factor increases the number of candidates for pivotal elements and thus the possibility of better preservation of the sparsity, i.e. again both storage and time may be reduced. If iterative refinement is used with large values of the above parameters then we accept that the computed LU-factorization will be inaccurate (hoping to compensate this by the iterations) but the storage for the non-zero elements and the time needed to perform the factorization tend to be reduced.

After this analysis we are able to give the following recommendations:

a). Iterative refinement may succesfully be used in cases (i), (iii) or (v). In these cases the matrix of the system must be factorized during each call of the package and therefore the time saved during the factorization may compensate for the time used to perform more forward and back substitutions in the iterative process.

b). If many systems with the same matrix are to be solved then iterative refinement is not efficient. It is better to use the double precision versions of the subroutines of the Y12M in this case.

c). If the matrix under consideration does not produce many fill-ins then the use of the direct option will be more efficient than the use of the iterative refinement option both with regard to the storage used and the computing time needed. However, if a reliable error estimation is required the use of the iterative refinement option should always be preferred.

1.3.6. Application of the Subroutines to Different Problems

We describe below the simplest way of using the package for each of the problems (i) - (v) given at the beginning of Section 1.1. Below we assume that iterative refinement will not be used.

(i) If only one system is to be solved then the simplest way to solve the system is by the use of subroutine Y12MA. The user must provide only the non-zero elements with their row and column numbers, the number of equations, the number Z of the non-zero elements and the right hand side vector. The length of arrays A, SNR and RNR should be about $2 \star Z$, i.e. $NN \geq 2 \star Z$ and $NN1 \geq 2 \star Z$ should also be initialized by the user.

(ii) If a sequence of systems with the same matrix is to be solved then subroutine Y12MA cannot be used efficiently, but a similar subroutine may be created in the following way: remove the statement IFLAG(5) = 1 from subroutine Y12MA and rename the new subroutine e.g. Y12MX. The user must initialize the same parameters as before the call of Y12MA plus IFLAG(5) = 2 (only NN should be larger) and call Y12MX when the first system of the sequence is solved. When the subsequent systems are solved only the right-hand side vector has to be initialized, IFLAG(5) should be set equal to 3 and subroutine Y12MD has to be called.

(iii) If a sequence of systems whose matrices are of the same structure is to be solved then Y12MA should be modified in the following way: remove the statement IFLAG(4) = 0 from subroutine Y12MA and rename the new subroutine e.g. Y12MY. When the first system of the sequence is solved initialize the same parameters as with the call of Y12MA plus IFLAG(4) = 1 and call Y12MY. When subsequent systems are solved initialize the same parameters as with the call of Y12MA plus IFLAG(4) = 2 and again call Y12MY. Remember that if the complete use of the fact that all matrices are of the same structure is required then larger values of NN and NN1 should be used.

(iv) If a sequence of systems whose matrices are of the same structure and some of them (but not all of them) are different is to be solved then subroutine Y12MA should be modified in the following way: remove the statements IFLAG(4) = 0 and IFLAG(5) = 1 from subroutine Y12MA and rename the new subroutine e.g. Y12MZ. When the first system of the sequence is solved initialize the same parameters as with the call of Y12MA plus IFLAG(4) = 1 and IFLAG(5) = 2 and call Y12MZ. If any of the subsequent systems whose matrix is the same as the matrix of the previous system is to be solved then initialize the new right-hand side vector, set IFLAG(5) = 3 and call Y12MD. If any of the subsequent systems whose matrix is different from the matrix of the previous system is to be solved then initialize the same parameters as with the call of Y12MA plus IFLAG(5) = 2 and IFLAG(4) = 2 and call Y12MZ. Remember that both NN and NN1 should be larger in this case (compared with the case where only one system is to be solved).

(v) If a sequence of systems whose matrices are of different structure is to be solved then subroutine Y12MA should be called to solve each system of the sequence and the same rules of initialization as in (i) are to be used before each call of subroutine Y12MA.

A subroutine similar to Y12MA may be written for the case where iterative refinement is used. Below we give only the values of the parameters (which are found experimentally): AFLAG(1) = 16, AFLAG(2) = a $\star$ 10^{-3} (where a is the magnitude of the elements of matrix A), AFLAG(3) = 10^{16}, AFLAG(4) = 10^{-12}, IFLAG(2) = 2 or 3, IFLAG(3) = 1, IFLAG(4) = 1, IFLAG(5) = 2 and IFLAG(11) = 16. In the subroutine corresponding to Y12MA (let us call it Y12MW) only Y12MF must be called (note that in Y12MA we call Y12MB, Y12MC and Y12MD). Similar rules apply when the new subroutine (Y12MW) is used to solve problems (i) - (v).

Note that in this subsection problems (i) - (v) are considered when the matrices of the systems are general. Sometimes special features of the matrix of the system (e.g. diagonal dominance) may be used to solve the system more efficiently (by the choice of the pivotal strategy, for example).

1.4. Numerical Results

The computations described in this section were performed on UNIVAC 1100/82 under the OS1100 Operating System previously called EXEC 8 at the Regional Computing Centre at the University of Copenhagen (RECKU). The single precision version of the subroutines was used and all computing times are given in seconds.

Matrices of two classes were used in the experiments. The matrices of these classes depend on two parameters: N - the order of the matrix and C - a parameter which determines the position of certain non-zero elements within the matrix. The sparsity pattern of the matrices of the first class, class D, is given in Fig. 1 (where the non-zero elements are denoted by #, the numerical values of the non-zero elements may be found in[17)18)). An example of a matrix of the second class, class E, is given in Fig. 2. The matrices of class D are general, while the matrices of class E are symmetric and positive definite.

We solved 48 systems with matrices of class D and 48 systems with matrices of class E (N = 250, 300, 350, 400, 450, 500, 550, 600 and C = 4, 44, 84, 124, 164, 204 for each N). The direct solutions (the solutions obtained without iterative refinement) and the solutions obtained with iterative refinement were compared (see Tables 1, 2, 3, and 4, where the computing time and the accuracy achieved are given; under "Error" we present the max-norm of x-y, where y is the computed solution and all components of the true solution, x, are equal to 1 for all systems solved).

The following conclusions can be drawn from the experimental results.

A) If iterative refinement is used more storage (compared with the storage needed when the system is solved directly) may sometimes be required (because when iterative refinement is

17.   Zlatev, Z. - "On Some Pivotal Strategies in Gaussian Elimination
        by Sparse Technique", SIAM J. Numer. Anal. 17, pp. 18-30, 1980.

18.   Zlatev, Z., Wasniewski, J., Schaumburg, K. - "A Testing Scheme
        For Subroutines Solving Large Linear Problems", Report No 81/1, The Regional
        Computing Centre at the University of Copenhagen, Vermundsgade 5,
        DK-2100 Copenhagen, Denmark, 1981 (to appear in Computers and
        Chemistry 5, 1981).

used we must store the non-zero elements of the lower triangular matrix L, we must make a copy of the original matrix and some other extra arrays are used). But note that if the system is large very often the accuracy requirements of the user can be satisfied only if double precision is used when the system is solved without iterative refinement. If this is so, then more storage will be needed when the system is solved directly.

B). Very often the computing time is reduced considerably when iterative refinement is used. In some of our examples the computing time is reduced to a third. When the number of equations in the systems increases the ratio of the computing time for the solution with iterative refinement and the computing time for the solution without iterative refinement tends to decrease (see Table 5, where the sums of the computing times for the solution of all systems with the same N, six for each N, are given).

	N=250	N=300	N=350	N=400	N=450	N=500	N=550	N=600
C=4								
time	0.64	0.77	0.85	0.96	1.12	1.15	1.31	1.44
error	0.0	0.0	0.0	0.0	0.0	0.0	0.0	0.0
C=44								
time	1.04	1.27	1.33	1.71	1.71	1.91	1.97	1.99
error	0.00	0.00	0.00	0.00	0.00	1.5E-8	0.00	0.00
C=84								
time	0.99	1.26	1.31	1.57	1.80	1.91	2.14	2.54
error	0.0	0.0	0.0	0.0	0.0	0.0	0.0	0.0
C=124								
time	0.68	1.33	1.43	1.52	1.66	1.75	1.95	2.42
error	0.0	0.0	0.0	0.0	0.0	0.0	0.0	0.0
C=164								
time	0.86	1.33	1.23	1.71	1.75	1.61	2.19	2.12
error	0.0	0.0	0.0	0.0	0.0	0.0	0.0	1.5E-8
C=204								
time	1.17	1.37	1.33	1.38	1.84	2.11	2.10	2.49
error	0.0	3.E-8	0.0	1.5E-8	0.0	0.0	0.0	3.E-8

Table 1 -   Systems whose matrices are of class D are solved by iterative refinement, with stability factor equal to 128.0 and drop-tolerance equal to 1.0E-3 (10^{-3}). Three rows with minimum numbers of non-zero elements are investigated in the pivotal search at each stage of the Gaussian elimination.

	N=250	N=300	N=350	N=400	N=450	N=500	N=550	N=600
C=4								
time	0.60	0.75	0.90	1.04	1.13	1.26	1.39	1.58
error	0.0	0.0	0.0	0.0	0.0	0.0	0.0	0.0
C=44								
time	1.07	1.40	1.89	2.30	2.87	3.32	4.10	4.83
error	0.0	1.5E-8	1.3E-7	1.5E-8	6.0E-8	1.5E-7	6.0E-8	1.5E-7
C=84								
time	0.68	0.92	1.12	1.50	1.92	2.26	2.69	2.98
error	0.0	0.0	1.5E-8	1.5E-8	0.0	0.0	1.5E-8	3.0E-8
C=124								
time	0.55	0.57	0.72	1.08	1.33	1.66	1.84	2.30
error	0.0	0.0	0.0	1.5E-8	0.00	1.5E-8	1.5E-8	0.0
C=164								
time	0.34	0.46	0.73	0.83	0.92	1.48	1.71	1.84
error	0.0	0.0	0.0	0.0	0.0	0.0	0.0	0.0
C=204								
time	0.30	0.37	0.48	0.57	0.94	1.02	1.10	1.20
error	0.0	0.0	0.0	0.0	0.0	0.0	0.0	0.0

Table 2 -   Systems whose matrices are of class E are solved by iterative refinement with the same parameters as in Table 1.

	N=250	N=300	N=350	N=400	N=450	N=500	N=550	N=600
C=4								
time	0.69	0.81	0.95	1.10	1.18	1.36	1.49	1.63
error	1.0E-4	5.2E-4	1.0E-4	2.4E-5	3.1E-5	9.3E-5	1.5E-4	3.8E-5
C=44								
time	2.42	3.27	2.89	4.69	5.77	7.10	6.90	7.68
error	5.7E-5	8.8E-5	5.9E-6	1.9E04	2.0E-4	3.7E-4	5.4E-5	2.9E-5
C=84								
time	1.21	2.99	2.71	4.79	5.05	5.07	7.10	7.01
error	5.9E-3	6.8E-4	2.8E-5	3.0E-5	3.5E-5	8.6E-5	2.3E-4	5.8E-5
C=124								
time	0.69	3.93	3.80	4.23	5.72	2.53	8.13	6.74
error	4.4E-5	1.1E-3	1.2E-5	2.1E-4	1.5E-4	1.1E-4	3.8E-6	1.5E-4
C=164								
time	1.24	2.87	3.49	4.85	6.59	2.50	7.78	8.66
error	2.9E-4	9.2E-4	1.2E-4	1.8E-4	4.5E-4	5.6E-3	1.9E-4	3.3E-4
C=204								
time	2.58	2.56	4.16	2.45	5.82	5.55	7.07	4.80
error	3.6E-4	6.0E-3	1.1E-4	1.2E-4	4.6E-3	7.9E-4	1.8E-4	2.2E-4

Table 3 -   Systems whose matrices are of class D are solved without iterative refinement, with stability factor equal to 16.0 and drop-tolerance equal to 1.0E-12 (10^{-12}). The non-zero elements of matrix L are not kept. Three rows with minimum numbers of non-zero elements are investigated in the pivotal search at each stage of the Gaussian elimination.

	N=250	N=300	N=350	N=400	N=450	N=500	N=550	N=600
C=4								
time	0.55	0.65	0.79	0.90	0.99	1.13	1.20	1.34
error	3.5E-5	4.9E-5	6.6E-5	8.6E-5	1.1E-4	1.3E-4	1.6E-4	1.9E-4
C=44								
time	1.27	1.74	2.47	3.38	4.24	5.32	6.38	8.46
error	1.6E-6	2.7E-6	2.7E-6	3.6E-6	4,8E-6	6.3E-6	7.2E-6	9.2E-6
C=84								
time	0.68	0.90	1.26	1.79	2.21	2.81	3.41	4.17
error	4.2E-7	5.7E-7	7.9E-7	1.0E-6	1.2E-6	1.9E-6	1.8E-6	2.5E-6
C=124								
time	0.34	0.58	0.85	1.14	1.27	1.80	2.03	2.76
error	2.2E-7	2.4E-7	3.9E-7	5.5E-7	5.4E-7	8.9E-7	8.8E-7	1.5E-6
C=164								
time	0.25	0.34	0.65	0.75	1.00	1.45	1.56	1.71
error	1.3E-7	1.2E-7	2.4E-7	2.4E-7	3.1E-7	5.8E-7	4.9E-7	4.5E-7
C=204								
time	0.22	0.28	0.36	0.55	0.84	0.98	1.09	I.57
error	I.3E-7	1.2E-7	1.3E-7	2.2E-7	2.4E-7	2.4E-7	2.5E-7	4.6E-7

Table 4 - Systems whose matrices are of class E are solved without iterative refinement, the same parameters as in Table 3 are used.

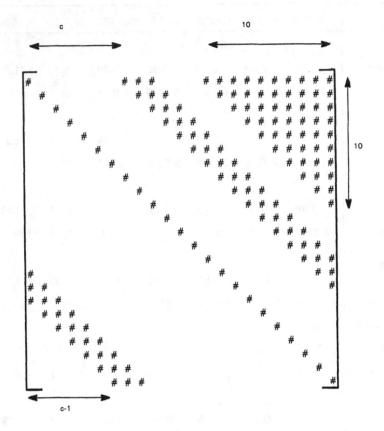

Figure 1

Sparsity pattern of a matrix of class D with N = 23 and C = 7.
(# denotes the non-zero elements)

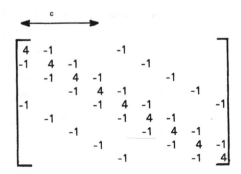

Figure 2

A matrix of class E with N = 9 and C = 4.

C) If single precision is used both when the systems are solved with iterative refinement and without iterative refinement, then the accuracy of the solution found with iterative refinement is better.

D) It is easy to compute the residual vector and to check the accuracy of the solution when iterative refinement is used. The last correction (stored on exit of Y12MF in array B) can also give some information about the accuracy of the solution. The subroutine computes an estimate of the error of the computed solution (stored in AFLAG(9)). Our experiments indicate that this estimate is very close to the true error. Therefore we can normally get a very good estimate of the error when iterative refinement is used. This is not possible when the solution is computed directly.

E). The value of the stability factor used with the package when iterative refinement is to be performed is not very critical. In our experiments we used 128.0 (see Table 1 and Table 2). Some experiments carried out by Zlatev[19] indicate that the use of small values of the stability factor (say, u=4.0) is more efficient when the drop-tolerance is large.

F). It is not easy to recommend values for the best drop-tolerance that can be used with all kinds of matrices. However, we believe that in each particular case the user will be able to find a good value for this parameter. In our experiments with matrices of class D and of class E (where the elements of the original matrix are in the interval from 0 to 1000) we have used 10^{-3} and this choice was quite satisfactory, but not optimal. It was the value which, in our opinion, was natural for our matrices. We find that the optimal value for the drop-tolerance is 10^{-2}. The experimental results are given in Table 6, where the total computing time is given for the cases when the solution is obtained without iterative refinement and with iterative refinement and different values of the drop-tolerance.

19.   Zlatev, Z. - "Use of Iterative Refinement in the Solution
        of Sparse Linear Systems", Report 1/79, Institute of Mathematics
        and Statistics, The Royal Veterinary and Agricultural University,
        Copenhagen, Denmark, 1979 (to appear in SIAM J. Numer. Anal.).

	WITHOUT ITERATIVE REFINEMENT		WITH ITERATIVE REFINEMENT	
N	CLASS D	CLASS E	CLASS D	CLASS E
250	8.83	3.45	5.38 (60.9%)	3.54(102.6%)
300	16.52	4.49	7.33 (44.4%)	4.47 (99.6%)
350	19.00	6.38	7.48 (39.4%)	5.84 (91.5%)
400	22.11	8.51	8.85 (40.0%)	7.32 (86.0%)
450	30.13	10.55	9.88 (32.8%)	9.11 (86.4%)
500	24.11	13.49	10.44 (43.3%)	11.00 (81.5%)
550	38.47	15.67	11.66 (30.3%)	12.83 (81.9%)
600	36.52	20.01	13.00 (35.6%)	14.73 (73.6%)

Table 5 -   Comparison of the computing times with and without iterative refinement.

G). We used the same length for the main arrays (A, SNR, RNR) in all experiments described in Tables 1 - 6. If systems whose matrices are of class E are solved with drop-tolerance equal to 10^{-4} (see Table 6) the length is too small when systems with N = 600 are solved, therefore many collections are needed, e.g. when N = 600 and C = 44 the number of collections is 32 and the computing time 8.74 secs; compare this with 1 collection and computing time 3.59 secs for the case where the drop-tolerance is equal to 10^{-2}. Note that the extra time is caused not only because many more collections are performed, but the main reason is that the sparsity is preserved better when the value of the drop-tolerance is larger. We have 6790 non-zero elements in the LU decomposition when the drop-tolerance is 10^{-2}, while for the value of the drop-tolerance 10^{-4} the corresponding number is 11203.

Class	Without iterative refinement	With iterative refinement and different drop-tolerance		
	10^{-10}	10^{-4}	10^{-3}	10^{-2}
Class D	195.69	94.83(43.3%)	74.02(37.8%)	66.97(34.2%)
Class E	82.98	80.98(98,1%)	68.84(83.4%)	63.34(76.7%)

Table 6 -   The total computing time with different options and with different drop-tolerances.

1.5. Contents of the Package

Subroutine Y12MA solves sparse systems of linear algebraic equations directly (i.e. without iterative refinement). The simplest way to use the package is to call Y12MA (or slightly modified subroutines as explained in Section 1.3.6). In this subroutine we initialize some components of arrays AFLAG and IFLAG and call the subroutines Y12MB, Y12MC and Y12MD (see below). The values of these parameters are chosen after performing many numerical experiments and they should give good results. If it is not so we recommend that the user checks these parameters in the documentation of the subroutines and changes their values if necessary.

Subroutine Y12MB orders the non-zero elements of a sparse matrix A by rows (the elements are in arbitrary order within a row) and prepares a sparse sysem of linear algebraic equations Ax = b to be factorized by Y12MC and solved by Y12MD. If the matrix is to be scaled it is convenient to do so after the call of Y12MB.

Subroutine Y12MC factorizes matrix A into two triangular matrices and prepares the system Ax = b to be solved by subroutine Y12MD.

Subroutine Y12MD solves the system Ax = b (whose matrix was factorized by Y12MC).

Subroutine Y12MF solves large and sparse systems with real coefficients by the use of Gaussian elimination and sparse technique; iterative refinement is applied in order to improve the accuracy. Very often the use of iterative refinement combined with the use of large values of the drop-tolerance (stored in AFLAG(2)) and/or of the stability factor (stored in AFLAG(1)) leads to a reduction of the total computing time needed to solve the system.

Both single and double precision versions of subroutines Y12MA to Y12MD are available. The double precision versions are denoted Y12MAF, Y12MBF, Y12MCF, Y12MDF; the single precision versions are denoted Y12MAE, Y12MBE, Y12MCE and Y12MDE. Only a single precision version of Y12MF is available and it is denoted by Y12MFE. The codes are written for UNIVAC 1100/82. It is possible that some changes will be needed if the codes are used on another computer (e.g. on some IBM installations INTEGER★2 declarations would be more efficient).

A subroutine which calculates an estimate of the condition number of a matrix is also available. Several generators for test-matrices are attached to the package and can be very useful during different experiments with the subroutines. Finally, a subroutine which exploits the UNIVAC facility "Multibanking" have been developed. This subroutine is very efficient on the UNIVAC 1100 series computers.

The codes of these subroutines with full documentation are in RECKU Library and can be requested from J. Wasniewski. Advances in the theory, and the experience of users, may prompt alterations to these codes. Readers and/or users are invited to write to the authors concerning any changes they may advocate. The users are kindly requested to refer to the present publication in each case where subroutines from this package are applied.

1.6. General remarks

More details concerning the implementation of iterative refinement in connection with sparse matrix technique and a large drop-tolerance can be found in Zlatev[20] and Zlatev, Schaumburg and Wasniewski[21]

The subroutines of the package have been compared to some NAG Subroutines[22] (more precisely the subroutines F01BRF and F04AXF from the NAG library; see also Duff[23] and Duff and Reid[24] . A full description of the results obtained in this comparison is given in

20.   Zlatev, Z. - "Use of Iterative Refinement in the Solution
       of Sparse Linear Systems", Report 1/79, Institute of Mathematics
       and Statistics, The Royal Veterinary and Agricultural University,
       Copenhagen, Denmark, 1979 (to appear in SIAM J. Numer. Anal.).

21.   Zlatev,Z. , Schaumburg,K. and Wasniewski,J. - "Implementation
       of an Iterative Refinement Option in a Code for Large and
       Sparse Systems", Computers and Chemistry, 4, pp. 87-99, 1980.

22.   NAG Library Fortran Manual, Mark 7, vol. 3, 4,
       Numerical Algorithms Group, 7 Banbury Road, Oxford OX2 6NN,
       United Kingdom.

23.   Duff, I.S. - "MA28 - a Set of FORTRAN Subroutines for Sparse Unsymmetric
       Matrices", Report No. R8730, A.E.R.E.,Harwell, England, 1977.

24.   Duff, I.S. and Reid, J.K. - "Some Design Features of a Sparse
       Matrix Code", ACM Trans. Math. Software 5, pp. 18-35, 1979.

Zlatev, Wasniewski and Schaumburg[25] . Many other examples are given in Zlatev[26] and Zlatev, Schaumburg and Wasniewski[27] .

The efficiency of the use of a large drop-tolerance has been pointed out by Clasen[28] , Wolfe[29] , Tewarson[30] and Reid[31] . The first package with an option where a large drop-tolerance may be specified and an attempt to regain the accuracy lost by iterative refinement is carried out, is Y12M. Note that the combination: *"sparse matrix technique and large drop-tolerance plus iterative refinement"* should be only an option in the sparse package since if many systems with the same matrix are to be solved then the direct solution may be a better choice.

The subroutines of the package have been used to solve linear least squares problems by forming augmented matrices (this method for solving linear least-squares problems is described e.g. in Bjorck[32]).

The iterative refinement option (i.e. the use of subroutine Y12MF) has been extremely

25.   Zlatev, Z., Wasniewski, J. and Schaumburg, K. - "Comparison of
        Two Algorithms for Solving Large Linear Systems". Report No 80/9,
        Regional Computing Centre at the University of Copenhagen (RECKU),
        Vermundsgade 5, 2100 Copenhagen, Denmark, 1980.

26.   Zlatev, Z. - "Use of Iterative Refinement in the Solution
        of Sparse Linear Systems", Report 1/79, Institute of Mathematics
        and Statistics, The Royal Veterinary and Agricultural University,
        Copenhagen, Denmark, 1979 (to appear in SIAM J. Numer. Anal.).

27.   Zlatev, Z., Schaumburg, K. and Wasniewski, J. - "Implementation
        of an Iterative Refinement Option in a Code for Large and Sparse Systems".
        Computers and Chemistry, 4, pp. 87-99, 1980.

28.   Clasen, R.J. - "Techniques for Automatic Tolerance in Linear
        Programming", Comm. ACM 9, pp. 802-803, 1966.

29.   Wolfe, P. - "Error in the Solution of Linear Programming Problems",
        In: "Error in Digital Computation" (L.B.Rall, ed.), vol 2, pp. 271-284,
        Wiley, New York, 1965.

30.   Tewarson, R.P. - "Sparse Matrices", Academic Press, New York, 1973.

31.   Reid, J.K. - "Fortran Subroutines for Handling Sparse Linear
        Programming Bases", Report R8269, A.E.R.E., Harwell, England, 1976.

32.   Bjorck, A. - "Methods for Sparse Linear Least-Squares Problems",
        In: "Sparse Matrix Computations" (J.Bunch and D.Rose, eds.),
        pp.177-199. Academic Press, New York, 1976.

efficient for such problems; see Zlatev[33] and Zlatev, Wasniewski and Schaumburg[34] . Some experiments with band, positive definite and symmetric matrices have also been carried out. The subroutines of package Y12M have been compared to a special routine for solving systems with such matrices (the subroutine F04ACF from the NAG Library[35]) see also Wilkinson and Reinsch[36] . The performance of our iterative refinement option has been comparable to the performance of the special subroutine F04ACF when the number of equations was larger than 900. If the system was very large (the number of equations larger than 1600) then the iterative refinement option of package Y12M performed even better. See more details for this experiment in Zlatev, Wasniewski and Schaumburg[34].

The subroutines of the package have been applied in the solution of some chemical problems arising in the theory of nuclear magnetic resonance spectroscopy. These problems can be described mathematically by large systems of **linear** ordinary differential equations, see Schaumburg and Wasniewski[37] Schaumburg, Wasniewski and Zlatev[38] and Zlatev, Wasniewski and Schaumburg[39] .

After time discretation by modified diagonally implicit Runge-Kutta methods (see

33.   Zlatev, Z. - "Use of Iterative Refinement in the Solution of Sparse Linear Systems", Report 1/79, Institute of Mathematics and Statistics, The Royal Veterinary and Agricultural University, Copenhagen, Denmark, 1979 (to appear in SIAM J. Numer. Anal.).

34.   Zlatev, Z., Wasniewski, J. and Schaumburg, K. - "Comparison of Two Algorithms for Solving Large Linear Systems". Report No 80/9, Regional Computing Centre at the University of Copenhagen (RECKU), Vermundsgade 5, 2100 Copenhagen, Denmark, 1980.

35.   NAG Library Fortran Manual, Mark 7, vol. 3, 4, Numerical Algorithms Group, 7 Banbury Road, Oxford OX2 6NN, United Kingdom.

36.   Wilkinson, J.H and Reinsch, C. - "Handbook for Automatic Computation", vol II, Linear Algebra, pp.50-56, Springer, Heidelberg, 1971.

37.   Schaumburg, K. and Wasniewski, J. - "Use of a Semiexplicit Runge-Kutta Integration Algorithm in a Spectroscopic Problem" Computers and Chemistry 2, pp. 19-25, 1978.

38.   Schaumburg, K., Wasniewski, J. and Zlatev, Z. - "Solution of Ordinary Differential Equations. Development of a Semiexplicit Runge-Kutta Algorithm and Application to a Spectroscopic Problem", Computers and Chemistry, 3, pp. 57-63, 1979.

39.   Zlatev, Z., Wasniewski, J. and Schaumburg, K. - "Classification of the Systems of Ordinary Differential Equations and Practical Aspects in the Numerical Integration of Large Systems", Computers and Chemistry, 4, pp. 13-18, 1980.

Schaumburg, Wasniewski and Zlatev[40] or Zlatev[41]) large sequences of linear systems of algebraic equations can be obtained. The subroutines of our package have succesfully been applied in the solution of these linear algebraic systems. Both the storage needed and the computation time used have been reduced considerably when the iterative refinement option with a large drop-tolerance has been used instead of solving the systems directly with the use of a small drop-tolerance. In some experiments the computing time has been reduced by more than a factor three. An algorithm for finding automatically the optimal drop-tolerance has been developed. This algorithm can always be used when large sequences of linear systems of algebraic equations are solved. See more details about these experiments in Schaumburg, Wasniewski and Zlatev[42] .

For the users at RECKU a version of the iterative refinement subroutine, that exploits the Univac facility: "Multibanking", has been developed. In this version two banks are used for the arrays. In the first bank all arrays which are used in the factorization and in the back substitution are kept (i.e. A, SNR, RNR, the first 11 columns of HA, PIVOT and B). In the the second bank we keep the arrays used in the computation of residual vectors $r_i = b - Ax_i$ (i.e. A1, SN, the 12th and 13th columns of HA, X and B1). Using only one bank at each stage of the computations (either the first bank or the second one) this version will never need more storage than the storage needed for the direct solution. The price is only a modest increase in the computing time. Thus this version combines the best features of Y12MA and Y12MF. It should be mentioned that this version is machine dependent, see more details in Wasniewski et al.[43]

40.   Schaumburg, K., Wasniewski, J. and Zlatev, Z. - "Solution
      of Ordinary Differential Equations. Development of a
      Semiexplicit Runge-Kutta Algorithm and Application to
      Spectroscopic Problem", Computers and Chemistry, 3, pp. 57-63, 1979.

41.   Zlatev, Z. - "Modified Diagonally Implicit Runge-Kutta
      Methods", Report No. 112, Department of Computer Science,
      University of Aarhus, Aarhus, Denmark, 1980 (to appear in
      SIAM Journal on Scientific and Statistical Computing).

42.   Schaumburg, K., Wasniewski, J. and Zlatev, Z. - "The Use of
      Sparse Matrix Technique in the Numerical Integration of Stiff
      Systems of Linear Ordinary Differential Equations". Computers and
      Chemistry, 4, pp. 1-12, 1980.

43.   Wasniewski, J., Zlatev, Z. and Schaumburg, K. -
      "A Multibanking Option of an Iterative Refinement Subroutine".
      In: "Conference Proceedings and Technical Papers",
      Spring Conference of Univac Users Assotiation/Europe, Geneva,
      1981.

An "ancestor" of package Y12M (subroutine SSLEST; see Zlatev, Barker and Thomsen[44])

has been used in the sparse solver for **non-linear** systems of ordinary differential equations

developed by Houbak and Thomsen[45] (see also Thomsen[46]).

Another "ancestor" of package Y12M (subroutine ST; see Zlatev and Thomsen[47]) has been

used in the solution of some parabolic partial differential equations by the algorithm

described in Zlatev and Thomsen[48][49] .

Very often one is interested in calculating the condition number of the matrix. The condition

number of the matrix is a measure of the sensivity of the solution to errors in the data and the

computation. Therefore the condition number may give useful information about the

accuracy lost during the calculations when the system is solved directly. The condition

number may also be useful when iterative refinement is used. This can be explained as

follows. Let COND be the condition number of matrix A and ε be the machine accuracy then

it is well known that the iterative refinement process is convergent if COND$\star\varepsilon$ has

44. Zlatev, Z., Barker, V.A. and Thomsen, P.G. -
"SSLEST - a FORTRAN IV Subroutine for Solving Sparse Systems
of Linear Equations (USER's GUIDE)", Report 78-01,
Institute for Numerical Analysis, Technical
University of Denmark, Lyngby, Denmark, 1978.

45. Houbak, N. and Thomsen, P.G. - "SPARKS - a FORTRAN
Subroutine for Solution of Large System of Stiff ODE's
with Sparse Jacobians", Report 79-02, Institute for
Numerical Analysis, Technical University of Denmark,
Lyngby, Denmark, 1979.

46. Thomsen, P.G. - "Numerical Solution of Large Systems
with Sparse Jacobians". In: "Working Papers for the
1979 SIGNUM Meeting on Numerical Ordinary Differential
Equations" (R.D. Skeel, ed.), University of Illinois at
Urbana - Champaign, Urbana, Illinois, 1979.

47. Zlatev, Z. and Thomsen., P.G. - "ST - a FORTRAN IV
Subroutine for the Solution of Large Systems of Linear
Algebraic Equations with Real Coefficients by Use of Sparse
Technique", Report 76-05, Institute for Numerical
Analysis, Technical University of Denmark, Lyngby,
Denmark, 1976.

48. Zlatev, Z. and Thomsen, P.G. - "Application of
Backward Differentiation Methods to the Finite
Element Solution of Time Dependent Problems",
International Journal for Numerical Methods
in Engineering, 14, pp. 1051 - 1061, 1979.

49. Zlatev, Z. and Thomsen, P.G. - "An Algorithm for the
Solution of Parabolic Partial Differential Equations
Based on Finite Element Discretization", Report 77-09,
Institute for Numerical Analysis, Technical
University of Denmark, Lyngby, Denmark, 1977.

magnitude at most comparable to 1. By the use of a drop-tolerance T we effectively replace the machine accuracy ε by T. This means, roughly speaking, that one can expect the iterative refinement process to be convergent if the magnitude of COND$\star$T is approximately equal to 1. Thus if one estimates COND, then it is possible to choose for the future runs T equal to (COND)$^{-1}$ and moreover one can expect this choice to be good.

Some very cheap and very efficient algorithms for evaluation of the condition number of a matrix are described in Forsythe et al.[50] , Cline et al.[51] and Moler[52] . Implementations for dense matrices can be found in Forsythe et al.[50] and in Dongarra et al.[53] An implementation for sparse matrices has been developed at RECKU (The Regional Computing Centre at the University of Copenhagen). This implementation, subroutine Y12MG, has successfully been tested on some matrices of class D and class E. The subroutine can optionally be called immediately after the factorization stage (after the subroutine Y12MC) without any special preparation. The use of the subroutine is very cheap, its cost is approximately equal to the computational cost of two back substitutions.

Finally, it is necessary to emphasize that the combination: "*sparse matrix technique and large drop tolerance plus iterative refinement*" can be used not only with the Gaussian

50. Forsythe, G.E., Malcolm, M.A., and Moler, C.B. - "Computer Methods for Mathematical Computations", Prentice-Hall, Englewood Cliffs, N.J., 1977.

51. Cline, A.K., Moler, C.B., Stewart, G.W. and Wilkinson, J.H. - "An Estimate for the Condition Number of a Matrix", SIAM J. Numer. Anal. 16, 368-375, 1979.

52. Moler, C.B. - "Three Research Problems in Numerical Linear Algebra". In: "Proceedings of the Symposia in Applied Mathematics" (G.H. Golub and J. Ollger, eds.) Vol. 22, pp. 1-18, American Mathematical Society, Providence, Rhode Island, 1978.

53. Dongarra, J.J., Bunch, J.R., Moler, C.B. and Stewart, G.W. - "LINPACK User's Guide", SIAM, Philadelphia, 1979.

decomposition but also with some other methods; see Zlatev[54)55)] and Zlatev and Nielsen[56)] .

54.   Zlatev, Z. - "On Solving Some Large Linear Problems
       by Direct Methods", Report 111, Department of Computer Science,
       University of Aarhus, Denmark, 1980.

55.   Zlatev, Z. - "Comparison of Two Pivotal Strategies in Sparse
       Plane Rotations", Report 122, Department of Computer Science,
       University of Aarhus, Aarhus, Denmark, 1980
       (to appear in Computers and Mathematics with Applications).

56.   Zlatev, Z. and Nielsen, H.B. - "Least - Squares
       Solution of Large Linear Problems". In:
       "Symposium i Anvendt Statistik 1980"
       (A. Hoskuldsson, K. Conradsen, B. Sloth Jensen and
       K. Esbensen, eds.), pp. 17 - 52, NEUCC, Technical
       University of Denmark, Lyngby, Denmark, 1980.

2. Documentation of subroutine Y12MA

2.1. Purpose

Y12MA solves sparse systems of linear algebraic equations by Gaussian elimination. The subroutine is a "black box subroutine" designed to solve efficiently problems which contain only one system with a single right hand side. The number of the input parameters is minimized. The user must assign values only to NN, NN1, N, Z, A, SNR, RNR, IHA and B according to the rules described in Section 2.4 (see below). It is extremely easy to modify the subroutine to the cases: **(a)** a sequence of systems with the same matrix is to be solved (note that one system with many right hand sides can be rewritten as a sequence of systems with the same matrix), **(b)** a sequence of systems whose matrices are different but of the same structure is to be solved and **(c)** a sequence of systems whose matrices are of the same structure and some of them are the same is to be solved. These cases are defined as case (ii), case (iii) and case (iv) in Section 1.1. "Scope of the Y12M". The recommendations in Section 1.3.6 should be followed in order to modify the subroutine Y12MA for the above cases. If a sequence of systems whose matrices are of different structure (this case appears as case (v) in Section 1.1) is to be solved then subroutine Y12MA should be called in the solution of each system in the sequence.

2.2. Calling sequence and declaration of the parameters

The subroutine is written in FORTRAN and it has been extensively tested with the FOR and FTN compilers on a UNIVAC 1100/82 computer, at the Regional Computing Centre at the

University of Copenhagen (RECKU). Many examples have been run on an IBM 3033 computer at the Northern Europe University Computing Centre (NEUCC) and on a CDC Cyber 173 computer at the Regional Computing Centre at the University of Aarhus (RECAU). Two different versions are available: a single precision version named Y12MAE and a double precision version named Y12MAF. The calls of these two versions and the declarations of the parameters are as follows.

A). Single precision version: Y12MAE

```
SUBROUTINE Y12MAE(N, Z, A, SNR, NN, RNR, NN1, PIVOT,
1                 HA, IHA, AFLAG, IFLAG, B, IFAIL)
REAL A(NN), PIVOT(N), B(N), AFLAG(8)
INTEGER SNR(NN), RNR(NN1), HA(IHA,11), IFLAG(10)
INTEGER N, Z, NN, NN1, IHA, IFAIL
```

B). Double precision version: Y12MAF

```
SUBROUTINE Y12MAF(N, Z, A, SNR, NN, RNR, NN1, PIVOT,
1                 HA, IHA, AFLAG, IFLAG, B, IFAIL)
DOUBLE PRECISION A(NN), PIVOT(N), B(N), AFLAG(8)
INTEGER SNR(NN), RNR(NN1), HA(IHA,11), IFLAG(10)
INTEGER N, Z, NN, NN1, IHA, IFAIL
```

These two versions can be used on many other computers also. However some alterations may be needed and/or may ensure greater efficiency of the performance of the subroutine. For example, it will be much more efficient to declare arrays SNR, RNR and (if possible) HA as INTEGER★2 arrays on some IBM installations.

2.3. Method

The system $Ax = b$ is solved by Gaussian elimination. Pivotal interchanges are used in an attempt to preserve both the stability of the computations and the sparsity of the original matrix. In this way a decomposition $LU = PAQ$ is normally calculated. P and Q are permutation matrices, L is a lower triangular matrix, U is an upper triangular matrix. The right hand side vector b is also modified during the decomposition so that the vector $c = L^{-1}Pb$ is available after the decomposition stage. In this way there is no need to store the non-zero elements of matrix L. Therefore these elements are not stored. This leads to a reduction of the storage requirements (the length of arrays A and SNR can be decreased). An approximation to the solution vector is found by solving $UQ^Tx = c$. The subroutine calculates the rate at which the elements of the matrix grow during the decomposition (see parameter AFLAG(5) below) and the minimal in absolute value pivotal element (see parameter AFLAG(8) below). These two numbers can be used to decide whether the computed approximation is acceptable or not. Positive values of a special parameter IFAIL indicate that the subroutine is unable to solve the problem. The error diagnostics, given in Section 7, describe the probable cause.

2.4. Parameters of the subroutine

N   -   INTEGER On entry N must contain the number of equations in the system $Ax=b$. **Unchanged on exit.**

Z   -   INTEGER. On entry Z must contain the number of non-zero elements in the coefficient matrix A of the system $Ax = b$. **Unchanged on exit.**

A   -   REAL (in the single precision version Y12MAE) or DOUBLE PRECISION (in the double precision version Y12MAF) array of length NN (see below). On entry the first Z locations of array A must contain the non-zero elements of the coefficient matrix A of the system Ax = b. The order of the non-zero elements may be completely arbitrary. **The content of array A is modified by subroutine Y12MA.** On successful exit array A will contain the non-zero elements of the upper triangular matrix U (without the diagonal elements of matrix U which can be found in array PIVOT, see below).

SNR - INTEGER array of length NN (see below). On entry SNR(j), j = 1(1)Z, must contain the column number of the non-zero element stored in A(j). **The content of array SNR is modified by subroutine Y12MA.** On successful exit array SNR will contain the column numbers of the non-zero elements of the upper triangular matrix U (without the column numbers of the diagonal elements of matrix U).

NN - INTEGER. On entry NN must contain the length of arrays A and SNR. **Restriction: NN $\geq$ 2$\star$Z.** Recommended value: 2$\star$Z $\leq$ NN $\leq$ 3$\star$Z. **Unchanged on exit.**

RNR - INTEGER array of length NN1 (see below). On entry RNR(i), i = 1(1)Z, must contain the row number of the non-zero element stored in A(i). **The content of array RNR is modified by subroutine Y12MA.** On successful exit all components of array RNR will normally be zero.

NN1 - INTEGER. On entry NN1 must contain the length of the array RNR. **Restriction: NN1 $\geq$ Z.** Recommended value: 2$\star$Z $\leq$ NN1 $\leq$ 3$\star$Z. **Unchanged on exit.**

PIVOT - REAL (in the single precision version Y12MAE) or DOUBLE PRECISION (in the double precision version Y12MAF) array of length N. **The content of array PIVOT is modified by subroutine Y12MA.** On successful exit array PIVOT will contain the pivotal elements (the diagonal elements of matrix U). This means that a small element (or small elements) in array PIVOT on exit may indicate numerical singularity of the coefficient matrix A. Note that the smallest in absolute value element in array PIVOT is also stored in AFLAG(8), see below.

HA   -   INTEGER two-dimensional array. The length of the first dimension is IHA (see below). The length of the second dimension is 11. This array is used as a work space by subroutine Y12MA.

IHA   -   INTEGER. On entry IHA must contain the length of the first dimension of array HA. **Restriction: IHA** $\geq$ **N. Unchanged on exit.**

AFLAG -   REAL (in the single precision version Y12MAE) or DOUBLE PRECISION (in the double precision version Y12MAF) array of length 8. **The content of array AFLAG is modified by subroutine Y12MA.** The content of the components of this array can be described as follows.

       AFLAG(1) -   **Stability factor.** An element can be chosen as pivotal element only if this element is larger (in absolute value) than the absolute value of the largest element in its row divided by AFLAG(1). Subroutine Y12MA sets AFLAG(1) = 16. This value has been found satisfactory for a wide range of test-examples.

       AFLAG(2) -   **Drop-tolerance.** An element, which in the process of the computations becomes smaller (in absolute value) than the drop-tolerance, is removed from array A (and its column and row numbers are removed from arrays SNR and RNR). Subroutine Y12MA sets AFLAG(2) = 10^{-12}. This value has been found satisfactory for a wide range of test-matrices. By this choice it is assumed that the matrix is not too badly scaled and that the magnitude of the elements is 1.

       AFLAG(3) -   The subroutine will stop the computations when the growth factor (parameter AFLAG(5), see below) becomes larger than AFLAG(3). Our experiments show that if AFLAG(3) $>$ 10^{16} then the solution is normally quite wrong. Therefore subroutine

Y12MA sets $AFLAG(3) = 10^{16}$.

AFLAG(4) -   The subroutine will stop the computations when the absolute value of a current pivotal element is smaller than AFLAG(4)$\star$AFLAG(6) (where the absolute value of the largest (in absolute value) element of the original matrix is stored in AFLAG(6), see below). Our experiments show that $AFLAG(4)=10^{-12}$ will normally be a good choice. Therefore subroutine Y12MA sets $AFLAG(4)=10^{-12}$.

AFLAG(5) -   **Growth factor.** After each stage of the elimination subroutine Y12MA sets AFLAG(5) = AFLAG(7)/AFLAG(6) (the description of parameters AFLAG(6) and AFLAG(7) is given below). Large values of AFLAG(5) indicate that an appreciable error in the computed solution is possible. In an extreme case when AFLAG(5) becomes larger than 10^{16}, see parameter AFLAG(3), subroutine Y12MA will stop the computations in an attempt to prevent overflows.

AFLAG(6) -   Subroutine Y12MA sets AFLAG(6) equal to $\max(|a_{ij}|)$, $i = 1(1)N$, $j = 1(1)N$.

AFLAG(7) -   Subroutine Y12MA sets AFLAG(7) equal to $\max(|a_{ij}^{(s)}|)$, $1 \leq s \leq k$, after each step k, k=1(1)N-1, of the elimination.

AFLAG(8) -   Subroutine Y12MA sets AFLAG(8) equal to $\min(|a_{ii}^{(i)}|)$, i= 1(1)N. This means that the minimal pivotal element (in absolute value) will be stored in AFLAG(8) on successful exit. Small values of AFLAG(8) indicate numerical singularity of the original matrix. We advise the user to check this parameter on exit very carefully.

IFLAG -   INTEGER array of length 10. **The content of this array is modified by subroutine Y12MA.** The content of the components of this array can be described as follows.

    IFLAG(1) -   Subroutine Y12MA uses IFLAG(1) as a work space.

    IFLAG(2) -   Subroutine Y12MA sets IFLAG(2) = 3. This means that at each stage of the Gaussian elimination (except the last one) the pivotal search is carried out in the 3 rows which have least numbers of non-zero elements.

    IFLAG(3) -   Subroutine Y12MA sets IFLAG(3) = 1. This means that the pivotal strategy for general matrices will be used. For some special matrices (for example positive definite) this will be inefficient. Subroutine Y12MA can easily be modified for such matrices; only the statement IFLAG(3) = 1 should be changed (for positive definite matrices e.g. IFLAG(3) = 2 can be used). More details about the use of this parameter with special matrices are given in Section 1.3.1

    IFLAG(4) -   Subroutine Y12MA sets IFLAG(4) = 0. This is the best choice in the case where only one system Ax = b is to be solved, see more details in Section 1.3.

    IFLAG(5) -   Subroutine Y12MA sets IFLAG(5) = 1. This means that the non-zero elements of the lower triangular matrix L will be removed during the decomposition stage, see more details in Section 1.3.3.

    IFLAG(6) -   On successful exit IFLAG(6) will be equal to the number of "garbage" collections in the row ordered list. If IFLAG(6) is large then it is better to choose a larger value of NN in the next calls of subroutine Y12MA with the same or a similar matrix A. This will lead to a reduction in the computing time.

IFLAG(7) -   On successful exit IFLAG(7) will be equal to the number of "garbage" collections in the column ordered list. If IFLAG(7) is large then it is better to choose a larger value of NN1 in the next calls of subroutine Y12MA with the same or a similar matrix A. This will lead to a reduction in the computing time.

IFLAG(8) -   On successful exit IFLAG(8) will be equal to the maximal number of non-zero elements kept in array A at any stage of the Gaussian elimination. If IFLAG(8) is much smaller then NN (or NN1) then the length NN (or NN1) can be chosen smaller in the next calls of subroutine Y12MA with the same or a similar matrix A. This will lead to a reduction of the storage needed.

IFLAG(9) -   This parameter is ignored by subroutine Y12MA. It will be used in some of the other subroutines when IFLAG(4) = 1 is specified on entry.

IFLAG(10)-   This parameter is ignored by subroutine Y12MA. It will be used in some of the other subroutines when IFLAG(4) = 1 is specified on entry.

B   - REAL (in the single precision version Y12MAE) or DOUBLE PRECISION (in the double precision version Y12MAF) array of length N. On entry the right hand side vector b of the system Ax = b must be stored in array B. **The content of array B is modified by subroutine Y12MA**. On successful exit the computed solution vector will be stored in array B.

IFAIL -   **Error diagnostic parameter. The content of parameter IFAIL is modified by subroutine Y12MA.** On exit IFAIL = 0 if the subroutine has not detected any error. Positive values of IFAIL on exit show that some error has been detected by the subroutine. Many of the error diagnostics are common for all subroutines in the package. Therefore the error diagnostics are listed in a separate section, Section 7, of this book. We advise the user to check the value of this parameter on exit.

2.5. Error diagnostics

Error diagnostics are given by positive values of the parameter IFAIL (see above). We advise the user to check carefully the value of this parameter on exit. The error messages are listed in Section 7.

2.6. Auxiliary subroutines

Y12MA calls three other subroutines: Y12MB, Y12MC and Y12MD.

2.7. Timing

The time taken depends on the order of the matrix (parameter N), the number of the non-zero elements in the matrix (parameter Z), the magnitude of the non-zero elements and their distribution in the matrix.

2.8. Storage

There are no internally declared arrays.

2.9. Accuracy

It is difficult to evaluate the accuracy of the computed solution. Large values of parameter AFLAG(5), **the growth factor**, indicate unstable computations during the decomposition stage. Small values of parameter AFLAG(8) can be considered as a signal for numerical singularity. We must emphasize here that normally much more reliable evaluations of the accuracy achieved can be found by the use of iterative refinement, i.e. by the use of subroutine Y12MF. By the use of the latter subroutine the computing time will often be reduced too. However, the storage requirements may be increased sometimes.

2.10. Some remarks

Remark 1   Subroutine Y12MA may also be considered as an example of how to initialize the first 4 components of array AFLAG and the first 5 components of array IFLAG when only one system with a single right hand side is to be solved. An efficient use of the subroutines Y12MB, Y12MC and Y12MD for the other cases discussed in Section 1.1 may be achieved by a single modification of Y12MA according to the rules given in Section 1.3.6 (a change of only one or two statements in Y12MA is needed to obtain such a modification).

Remark 2   The last 4 components of array AFLAG and the last 5 components of array IFLAG can be used as output parameters. The user is not obliged to do this. However, we recommend the check of these parameters on exit.

2.11. Example.

Consider the following system:

$$Ax = b,$$

where

$$A = \begin{matrix}
10 & 0 & 0 & 0 & 0 & 0 \\
0 & 12 & -3 & -1 & 0 & 0 \\
0 & 0 & 15 & 0 & 0 & 0 \\
-2 & 0 & 0 & 20 & -2 & 0 \\
-1 & 0 & 0 & -5 & 1 & -1 \\
-1 & -2 & 0 & 0 & 0 & 6
\end{matrix}$$

and

$$b = (10, 11, 45, 33, -22, 31)^T.$$

The above system is solved by the single precision version Y12MAE.

2.11.1. Program

```
       PARAMETER IHA = 10,NN = 200,NN1 = 100
       IMPLICIT REAL (A-B,G,P,T-V),INTEGER(C,F,H-N,R-S,Z)
       REAL A(NN),PIVOT(IHA),B(IHA),AFLAG(8)
       INTEGER SNR(NN),RNR(NN1),HA(IHA,11),IFLAG(10)
       DATA NIN/5/, NOUT/6/
C
C      INITIALIZATION OF THE PARAMETERS.
```

```
C

      READ (NIN,101)N,Z
101   FORMAT(2I4)
C

C     INITIALIZE THE NON-ZERO ELEMENTS OF MATRIX
C     A IN ARBITRARY ORDER.
C

      DO 120 K = 1,Z
      READ(NIN,110) RNR(K), SNR(K), A(K)
110   FORMAT(2I4,F12.6)
120   CONTINUE
C

C

C     INITIALIZE THE COMPONENTS OF THE RIGHT-HAND SIDE VECTOR B.
C

      READ(NIN,130) (B(K),K = 1,N)
130   FORMAT(6F12.6)
C

C     CALL THE SUBROUTINE Y12MAE.
C

      CALL Y12MAE(N,Z,A,SNR,NN,RNR,NN1,PIVOT,HA,IHA,AFLAG,
     1 IFLAG,B,IFAIL)
C

C     PRINT THE RESULTS.
C

      WRITE(NOUT,1)IFAIL
1     FORMAT('1THE ERROR DIAGNOSTIC PARAMETER  IFAIL',
     1 ' IS EQUAL TO',I4)
      IF(IFAIL.GT.0)GO TO 5
      WRITE(NOUT,2)
2     FORMAT('0THE SOLUTION VECTOR IS GIVEN BELOW.')
```

```
        DO 4 I = 1,6
        WRITE(NOUT,3)I,B(I)
3       FORMAT(' ',I10,F20.5)
4       CONTINUE
C
C       PRINT THE AUXILIARY INFORMATION ABOUT THE SOLUTION.
C       THIS IS OPTIONALLY.
C
        WRITE(NOUT,11)AFLAG(6)
11      FORMAT('0THE LARGEST ELEMENT IN THE ORIGINAL',
       1 ' MATRIX IS:'F12.5)
        WRITE(NOUT,12)AFLAG(7)
12      FORMAT('0THE LARGEST ELEMENT FOUND IN THE ELIMINATION',
       1 F12.5)
        WRITE(NOUT,13)AFLAG(8)
13      FORMAT('0THE MINIMAL(IN ABSOLUTE VALUE)PIVOTAL ELEMENT',
       1 F12.5)
        WRITE(NOUT,14)AFLAG(5)
14      FORMAT('0THE GROWTH FACTOR IS:',1PD12.2)
        WRITE(NOUT,15)IFLAG(6)
15      FORMAT('0THE NUMBER OF COLLECTIONS IN THE ROW LIST',I5)
        WRITE(NOUT,16)IFLAG(7)
16      FORMAT('0THE NUMBER OF COLLECTIONS IN THE COLUMN LIST',I5)
        WRITE(NOUT,17)IFLAG(8)
17      FORMAT('0THE LARGEST NUMBER OF ELEMENTS FOUND IN ARRAY A',
       1 I9)
5       CONTINUE
        STOP
        END
```

2.11.2. Input

```
6  15
1   1  10.
6   6   6.
6   2  -2.
6   1  -1.
2   2  12.
2   3  -3.
2   4  -1.
4   1  -2.
5   1  -1.
5   6  -1.
5   5   1.
5   4  -5.
4   4  10.
4   5  -1.
3   3  15.
10.      11.      45.      33.      -22.      31.
```

2.11.3. Output

```
THE ERROR DIAGNOSTIC PARAMETER IFAIL IS EQUAL TO   0
THE SOLUTION VECTOR IS GIVEN BELOW.
        1          1.00000
        2          2.00000
        3          3.00000
        4          4.00000
        5          5.00000
        6          6.00000
THE LARGEST ELEMENT IN THE ORIGINAL MATRIX IS:   15.00000
THE LARGEST ELEMENT FOUND IN THE ELIMINATION   15.00000
THE MINIMAL(IN ABSOLUTE VALUE)PIVOTAL ELEMENT      .49722
THE GROWTH FACTOR IS:   1.00+000
THE NUMBER OF COLLECTIONS IN THE ROW LIST   0
THE NUMBER OF COLLECTIONS IN THE COLUMN LIST   0
THE LARGEST NUMBER OF ELEMENTS FOUND IN ARRAY A      15
```

3. Documentation of subroutine Y12MB

3.1. Purpose

Y12MB prepares a system of linear algebraic equations to be factorized (by subroutine Y12MC) and solved (by subroutine Y12MD). Each call of subroutine Y12MC must be preceded by a call of Y12MB. It is very convenient to perform some operations on the non-zero elements of the coefficient matrix between calls of Y12MB and Y12MC. For example, when iterative refinement is carried out the information needed in the calculation of the residual vector is copied in some extra arrays between the calls of Y12MB and Y12MC. It is also very easy to perform any kind of scaling after the call of Y12MB.

3.2. Calling sequence and declaration of the parameters

The subroutine is written in FORTRAN and has been extensively tested with the FOR and FTN compilers on a UNIVAC 1100/82 computer at the Regional Computing Centre at the University of Copenhagen (RECKU). Many examples have been run on an IBM 3033 computer at the Northern Europe University Computing Centre (NEUCC) and on a CDC Cyber 173 computer at the Regional Computing Centre at the University of Aarhus (RECAU). Two different versions are available: a single precision version Y12MBE and a double precision version Y12MBF. The calls of these two versions and the declaration of the parameters are as follows:

A) Single precision version Y12MBE

SUBROUTINE Y12MBE(N, Z, A, SNR, NN, RNR, NN1, HA, IHA,

1   AFLAG, IFLAG, IFAIL)

REAL A(NN), AFLAG(8)

INTEGER SNR(NN), RNR(NN1), HA(N,11), IFLAG(10)

INTEGER N, Z, IHA, IFAIL

B) Double precision version Y12MBF

SUBROUTINE Y12MBF(N, Z, A, SNR, NN, RNR, NN1, HA, IHA,

1   AFLAG, IFLAG, IFAIL)

DOUBLE PRECISION A(NN), AFLAG(8)

INTEGER SNR(NN), RNR(NN1), HA(N,11), IFLAG(10)

INTEGER N, Z, IHA, IFAIL

These two versions can be used on many other computers also. However some alterations may be needed and/or may ensure greater efficiency of the performance of the subroutine. For example, it will be much more efficient to declare arrays SNR, RNR and (if possible) HA as INTEGER★2 arrays on some IBM installations.

3.3. Method

The non-zero elements of matrix A are ordered by rows and stored in the first Z positions of array A. The order of the non-zero elements within a row is arbitrary. The column numbers of the non-zero elements are stored in the first Z positions of array SNR so that if $A(J)=a_{ij}$, $J=1(1)Z$, then $SNR(J)=j$. The row numbers of the non-zero elements are stored in the first Z positions of array RNR so that the row numbers of the non-zero elements in the first column of matrix A are located before the row numbers of the non-zero elements in the second column of matrix A, the row numbers of the non-zero elements in the second column of matrix A are located before the row numbers of the non-zero elements in the third column of matrix A and so on. Some additional information, e.g. about the row starts, row ends, column starts and column ends, is stored in the work array HA. **This storage scheme has been proposed by Gustavson[1].**

3.4. Parameters of the subroutine

N — INTEGER. On entry N must contain the number of equations in the system Ax=b. **Unchanged on exit.**

Z — INTEGER. On entry Z must contain the number of non-zero elements in the coefficient matrix A of the system Ax = b. **Unchanged on exit.**

A — REAL (in the single precision version Y12MBE) or DOUBLE PRECISION (in the double precision version Y12MBF) array of length NN (see below). On entry the first Z locations of array A must contain the non-zero elements of the coefficient matrix A of the system Ax = b. The order of the non-zero elements may be completely arbitrary. **The content of array A is modified by subroutine Y12MB.** On successful exit the first Z positions of array A will contain the non-zero elements of the matrix A ordered by rows.

SNR — INTEGER array of length NN (see below). On entry SNR(j), j = 1(1)Z, must contain the column number of the non-zero element stored in A(j). **The content of array SNR is modified by subroutine Y12MB.** On successful exit SNR(j), j=1(1)Z, will contain the column number of the non-zero element stored in A(j).

1.   Gustavson, F.G. - "Some Basic Techniques for Solving Sparse
      System of Linear Equations",
      In: "Sparse Matrices and Their Applications",
      (D.J. Rose and R.A. Willoughby, eds.), pp. 41-52,
      Plenum Press, New York, 1972.

SNR   -

NN   - INTEGER. On entry NN must contain the length of arrays A and SNR. **Restriction:**
NN $\geq$ 2$\star$Z. Recommended value: 2$\star$Z $\leq$ NN $\leq$ 3$\star$Z. See also the description of
NN in Y12MC. **Unchanged on exit.**

RNR   - INTEGER array of length NN1 (see below). On entry RNR(i), i = 1(1)Z, must
contain the row number of the non-zero element stored in A(i). **The content of**
array RNR is modified by subroutine Y12MB. On successful exit the row
numbers of the non-zero elements of matrix A are stored in the first Z positions
of array RNR, so that the row numbers of the non-zero elements in the first
column of matrix A are before the row numbers of the non-zero elements in the
second column of matrix A, the row numbers of the non-zero elements in the
second column of matrix A are before the row numbers of the non-zero elements
in the third column of matrix A and so on. This means that on exit the row
number of the non-zero element stored in A(i), i = 1(1)Z, is not stored in RNR(i),
in general.

NN1   - INTEGER. On entry NN1 must contain the length of the array RNR. **Restriction:**
NN1 $\geq$ Z. Recommended value: 2$\star$Z $\leq$ NN1 $\leq$ 3$\star$Z. See also the desription of
NN1 in Y12MC. **Unchanged on exit.**

HA   - INTEGER two-dimensional array. The length of the first dimension of HA is IHA
(see below). The length of the second dimension of HA is 11. **The contents of**
some elements of this array are modified by subroutine Y12MB, the
contents of the others are ignored. On successful exit:

    (1) HA(i,1)   contains the position in array A where the first element of row
                     i, i = 1(1)N, is stored.
    (2) HA(i,3)   contains the position in array A where the last element of row
                     i, i = 1(1)N, is stored (all non-zero elements of row i are
                     located between HA(i,1) and HA(i,3) compactly; the number of

non-zero elements in row i is HA(i,3) - HA(i,1) + 1).

(3) HA(i,4)   contains the position in array RNR where the row number of the first non-zero element of column j is stored (j = 1(1)N).

(4) HA(j,6)   contains the position in array RNR where the row number of the last element in column j, j = 1(1)N, is stored (all row numbers of the non-zero elements of column j are located between HA(j,4) and HA(j,6) compactly, the number of non-zero elements in column j is HA(j,6)-HA(j,4)+1).

(5)   Some information needed in the pivotal search during the decomposition is stored in columns 7,8 and 11 of array HA.

(6)   The other columns of HA are either ignored by Y12MB or used as a work space.

IHA - INTEGER. On entry IHA must contain the length of the first dimension of array HA. **Restriction: IHA $\geq$ N. Unchanged on exit.**

AFLAG - REAL (in the single precision version Y12MBE) or DOUBLE PRECISION (in the double precision version Y12MBF) array of length 8. **The contents of all locations of array AFLAG, except AFLAG(6), are ignored by subroutine Y12MB. The content of AFLAG(6) is modified by Y12MB.** On successful exit AFLAG(6) contains max($|a_{ij}|$).

IFLAG - INTEGER array of length 10. The user should set IFLAG(1) $\geq$ 0 before the call of package Y12M (i.e. before the first call of a subroutine of this package). IFLAG(1) is used in the error checks by the subroutines and should not be altered by the user between any two successive calls of subroutines of the package. IFLAG(1) will be equal to -1 on successful exit from subroutine Y12MB. Thus, **the content of IFLAG(1) is modified by subroutine Y12MB.** On entry IFLAG(4) must contain 0, 1 or 2. IFLAG(4) = 0 is the best choice when only one system is to be solved, when the first system of a sequence of systems with the same matrix is to be solved and when any system of a sequence of systems whose matrices are of different structure is to be solved. IFLAG(4) = 1 is the best choice when the first system in a sequence of systems with the same structure is to be solved. IFLAG(4) = 2 is the best choice when any system after the first one in a sequence of systems whose matrices are of the same structure is to be solved. **The content of IFLAG(4) is unchanged by subroutine Y12MB.** The other locations of IFLAG are either ignored by Y12MB or used as a work space.

IFAIL -   **Error diagnostic parameter. The content of parameter IFAIL is modified by subroutine Y12MB.** On exit IFAIL = 0 if the subroutine has not detected any error. Positive values of IFAIL on exit show that some error has been detected by the subroutine. Many of the error diagnostics are common for all subroutines in the package. Therefore the error diagnostics are listed in a separate section, Section 7, of this book. We advise the user to check the value of this parameter on exit.

3.5. Error diagnostics

Error diagnostics are given by positive values of the parameter IFAIL (see above). We advise the user to check carefully the value of this parameter on exit. The error messages are listed in Section 7.

3.6. Auxiliary subroutines

None.

3.7. Timing

The time taken depends on the number of non-zero elements (parameter Z) in matrix A.

3.8. Storage

There are no internally declared arrays.

3.9. Some remarks

Remark 1   The use of subroutine Y12MB is followed by the use of Y12MC and Y12MD. Subroutines Y12MA and Y12MF can be considered as examples of how to use Y12MB, Y12MC and Y12MD.

Remark 2   The contents of N, Z, A, SNR, NN, RNR, NN1, columns 1, 3, 4, 6, 7, 8 and 11 of HA, IHA, AFLAG(6), IFLAG(1), IFLAG(4) and IFAIL should not be altered between calls of Y12MB and Y12MC.

Remark 3   If IFAIL $>$ 0 on exit of Y12MB there is no sense in calling Y12MC and Y12MD. Therefore the computations should be stopped in this case and one should investigate (using the error diagnostics given in Section 7) why Y12MB has assigned a positive value to IFAIL.

3.10. Example.

Consider the following system:

$$Ax = b,$$

where

$$A = \begin{matrix} 10 & 0 & 0 & 0 & 0 & 0 \\ 0 & 12 & -3 & -1 & 0 & 0 \\ 0 & 0 & 15 & 0 & 0 & 0 \\ -2 & 0 & 0 & 20 & -2 & 0 \\ -1 & 0 & 0 & -5 & 1 & -1 \\ -1 & -2 & 0 & 0 & 0 & 6 \end{matrix}$$

and

$$b = (10, 11, 45, 33, -22, 31)^{\mathsf{T}}.$$

Input data - Some components of the arrays AFLAG and IFLAG must be initialized (see the code). We give below only the data initialized by READ statements.

Output data - The first group of numbers represents the position of the elements in the row ordered list, their column numbers and the non-zero elements themselves (e.g. the numbers given in the first line show that: at the first position of array A, there is a non-zero element with column number one and the element itself is 10.00000). The second group of numbers gives information about the row starts and row ends (e.g. the numbers given in the second line, under I HA(I,1) HA(I,3), show that: the non-zero elements of the second row begin at position two and end at position four). The third group of numbers gives information about the position of the non-zero elements (ordered by columns) and their row numbers. Note that the non-zero elements are not actually ordered by columns, thus e.g. the information written on the seventh line (under I RNR(I)) should be interpreted in the following way: if we order the non-zero elements by columns, then the seventh element would have row number two. The last group of numbers gives information about the column starts and the column ends (e.g. if the non-zero elements are ordered by columns then the numbers on the first line, under I HA(I,4) HA(I,6), show that the elements of the first column would begin at postition one and finish at position four).

The single precision version Y12MBE is used in the example.

3.10.1. Program

```
          PARAMETER IHA = 10,NN = 200,NN1 = 100
          IMPLICIT REAL(A-B,G,P,T-V),INTEGER(C,F,H-N,R-S,Z)
          REAL A(NN),AFLAG(8)
          INTEGER SNR(NN),RNR(NN1),HA(IHA,11),IFLAG(10)
          DATA NIN/5/, NOUT/6/
C
C         INITIALIZATION OF THE PARAMETERS.
C
          AFLAG(1) = 16.0
          AFLAG(2) = 1.E-12
          AFLAG(3) = 1.E+16
          AFLAG(4) = 1.E-12
          IFLAG(1) = 0
          IFLAG(2) = 3
          IFLAG(3) = 1
          IFLAG(4) = 0
          IFLAG(5) = 1
          READ (NIN,101)N,Z
101       FORMAT(2I4)
```

```
C
C
C       INITIALIZE THE NON-ZERO ELEMENTS OF MATRIX A
C       IN ARBITRARY ORDER.
C

        DO 120 K = 1,Z
        READ(NIN,110) RNR(K), SNR(K), A(K)
110     FORMAT(2I4,F12.6)
120     CONTINUE
C
C       CALL THE SUBROUTINE Y12MBE.
C

        CALL Y12MBE(N,Z,A,SNR,NN,RNR,NN1,HA,IHA,AFLAG,
       1 IFLAG,IFAIL)
C
C       PRINT THE RESULTS.
C

        WRITE(NOUT,1)IFAIL
1       FORMAT('1THE ERROR DIAGNOSTIC PARAMETER  IFAIL',
       1 ' IS EQUAL TO',I4)
        IF(IFAIL.GT.0)GO TO 5
        WRITE(NOUT,22)
22      FORMAT('0    I    SNR(I)      A(I)')
        DO 25 I = 1,Z
        WRITE(NOUT,26) I, SNR(I), A(I)
26      FORMAT(2I10,F20.5)
25      CONTINUE
        WRITE(NOUT,31)
31      FORMAT('0    I    HA(I,1)    HA(I,3)')
        DO 32 I = 1,N
        WRITE(NOUT,33) I, HA(I,1), HA(I,3)
```

```
33        FORMAT(3I11)

32        CONTINUE

          WRITE(NOUT,35)

35        FORMAT('0     I      RNR(I)')

          DO 37 I = 1,Z

          WRITE(NOUT,38) I, RNR(I)

38        FORMAT(2I11)

37        CONTINUE

          WRITE(NOUT,42)

42        FORMAT('0     I      HA(I,4)  HA(I,6)')

          DO 43 I = 1,N

          WRITE(NOUT,44) I, HA(I,4), HA(I,6)

44        FORMAT(3I11)

43        CONTINUE

5         CONTINUE

          STOP

          END
```

3.10.2. Input

```
6  15
1  1  10.
6  6   6.
6  2  -2.
6  1  -1.
2  2  12.
2  3  -3.
2  4  -1.
4  1  -2.
5  1  -1.
5  6  -1.
5  5   1.
5  4  -5.
4  4  10.
4  5  -1.
3  3  15.
```

3.10.3. Output

THE ERROR DIAGNOSTIC PARAMETER IFAIL IS EQUAL TO 0

I	SNR(I)	A(I)
1	1	10.00000
2	2	12.00000
3	3	-3.00000
4	4	-1.00000
5	3	15.00000
6	1	-2.00000
7	4	10.00000
8	5	-1.00000
9	1	-1.00000
10	6	-1.00000
11	5	1.00000
12	4	-5.00000
13	6	6.00000
14	2	-2.00000
15	1	-1.00000

I	HA(I,1)	HA(I,3)
1	1	1
2	2	4
3	5	5
4	6	8
5	9	12
6	13	15

I	RNR(I)
1	1
2	4
3	5
4	6
5	2
6	6
7	2
8	3
9	2
10	4
11	5
12	4
13	5
14	5
15	6

I	HA(I,4)	HA(I,6)
1	1	4
2	5	6
3	7	8
4	9	11
5	12	13
6	14	15

4. Documentation of subroutine Y12MC

4.1. Purpose

Y12MC decomposes a matrix A into two triangular matrices L and U. Each call of subroutine Y12MC should be preceded by a call of Y12MB. Subroutine Y12MD is normally called one or several times after the call of Y12MC.

4.2. Calling sequence and declaration of the parameters

The subroutine is written in FORTRAN and has been extensively tested with the FOR and FTN compilers on the UNIVAC 1100/82 computer at the Regional Computing Centre at the University of Copenhagen (RECKU). Many examples have been run on an IBM 3033 computer at the Northern Europe University Computing Centre (NEUCC) and on a CDC Cyber 173 computer at the Regional Computing Centre at the University of Aarhus (RECAU). Two different versions are available: a single precision version named Y12MCE and a double precision version named Y12MCF. The calls of these two versions and the declaration of the parameters are as follows.

A). Single precision version: Y12MCE

```
      SUBROUTINE Y12MCE(N, Z, A, SNR, NN, RNR, NN1, PIVOT,
     1                  B, HA, IHA, AFLAG, IFLAG, IFAIL)
      REAL A(NN), PIVOT(N), B(N), AFLAG(8)
      INTEGER SNR(NN), RNR(NN1), HA(IHA,11), IFLAG(10)
      INTEGER N, Z, NN, NN1, IHA, IFAIL
```

B). Double precision version: Y12MCF

```
      SUBROUTINE Y12MCF(N, Z, A, SNR, NN, RNR, NN1, PIVOT,
     1                  B, HA, IHA, AFLAG, IFLAG, IFAIL)
      DOUBLE PRECISION A(NN), PIVOT(N), B(N), AFLAG(8)
      INTEGER SNR(NN), RNR(NN1), HA(IHA,11), IFLAG(10)
      INTEGER N, Z, NN, NN1, IHA, IFAIL
```

These two versions can be used on many other computers also. However some alterations may be needed and/or may ensure greater efficiency of the performance of the subroutine. For example, it will be much more efficient to declare arrays SNR, RNR and (if possible) HA as INTEGER★2 arrays on some IBM installations.

4.3. Method

Gaussian elimination is used in the decomposition of matrix A. Pivotal interchanges are implemented as an attempt to preserve both the stability of the computations and the sparsity of the coefficient matrix. Two triangular matrices L and U are computed so that LU=PAQ (where P and Q are permutation matrices). The right hand side vector b is also modified so that vector $c = L^{-1}Pb$ is computed on successful exit. In this way matrix L is sometimes not needed in the further computations and may be removed if this is so.

4.4. Parameters of the subroutine

N   -   INTEGER. On entry N must contain the number of equations in the system Ax=b. **Unchanged on exit.**

Z   -   INTEGER. On entry Z must contain the number of non-zero elements in the coefficient matrix A of the system Ax=b. **Unchanged on exit.**

A   -   REAL (in the single precision version Y12MCE) or DOUBLE PRECISION (in the double precision version Y12MCF) array of length NN (see below). On entry the first Z locations of array A must contain the non-zero elements of the coefficient matrix A of the system Ax = b ordered by rows (in the preceding call of Y12MB). **The content of array A is modified by subroutine Y12MC.** On successful exit array A will contain the non-zero elements of the upper triangular matrix U (without the diagonal elements of matrix U which can be found in array PIVOT, see below) and sometimes also the non-zero elements of the lower triangular matrix L (without the diagonal elements which are not stored).

SNR   -   INTEGER array of length NN (see below). On entry SNR(j), j = 1(1)Z, must contain the column number of the non-zero element stored in A(j). This is accomplished by subroutine Y12MB. **The content of array SNR is modified by subroutine Y12MC.** On successful exit array SNR will contain the column numbers of the non-zero elements of the upper triangular matrix U (without the column numbers of the diagonal elements of matrix U) and sometimes also the column numbers of the non-zero elements of the lower triangular matrix L (without the column numbers of the diagonal elements of L).

NN   -   INTEGER. On entry NN must contain the length of array A and SNR. **Restriction:**
           NN ≥ 2⋆Z. Recommended value: 2⋆Z ≤ NN ≤ 3⋆Z if the non-zero elements of
           matrix L will be removed by subroutine Y12MC (i.e. if Y12MC is called with
           IFLAG(5) = 1) or 3⋆Z ≤ NN ≤ 5⋆Z if the non-zero elements of matrix L will be
           kept by subroutine Y12MC (i.e. if Y12MC is called with IFLAG(5) = 2).
           Unchanged on exit.

RNR   -   INTEGER array of length NN1 (see below). On entry the first Z locations of array
           RNR must contain the row numbers of the non-zero elements of the coefficient
           matrix A of the system Ax = b so that the row numbers of the non-zero elements
           in the first column of matrix A are located before the row numbers of the
           non-zero elements in the second column of matrix A, the row numbers of the
           non-zero elements in the second column of matrix A are located before the row
           numbers of the non-zero elements in the third column of matrix A and so on (this
           ordereing is prepared by subroutine Y12MB). **The content of array RNR is**
           modified by subroutine Y12MC. On successful exit all components of array
           RNR will normally be zero.

NN1   -   INTEGER. On entry NN1 must contain the length of the array RNR. **Restriction:**
           NN1 ≥ Z. Recommended value: **2⋆Z ≤ NN1 ≤ 3⋆Z. Unchanged on exit.**

PIVOT   -   REAL (in the single precision version Y12MCE) or DOUBLE PRECISION (in the
           double precision version Y12MCF) array of length N. **The content of array**
           PIVOT is modified by subroutine Y12MC. On successful exit array PIVOT will
           contain the pivotal elements (the diagonal elements of matrix U). This means
           that a small element (or small elements) in array PIVOT on exit may indicate
           numerical singularity of the coefficient matrix A. Note that the smallest in
           absolute value element in array PIVOT is stored in AFLAG(8), see below.

B   -   REAL (in the single precision version Y12MCE) or DOUBLE PRECISION (in the
           double precision version Y12MCF) array of length N. The right hand side vector b

of the system $Ax = b$ must be stored in B on entry. **The content of array B is modified by subroutine Y12MC.** On successful exit array B will contain the components of vector $c = L^{-1}Pb$.

HA   -   INTEGER two-dimensional array. The length of the first dimension of HA is IHA (see below). The length of the second dimension of HA is 11. The content of columns 1, 3, 4, 6, 7, 8 and 11 is prepared by Y12MB (and thus they should not be altered between calls of Y12MB and Y12MC). **The content of array HA is modified by subroutine Y12MC.** The non-zero elements (without the diagonal elements) of row i in matrix L (when this matrix is stored) are between $HA(i,1)$ and $HA(i,2)$ -1. If the non-zero elements of matrix L are not stored then $HA(i,1)=HA(i,2)$, $i=1(1)N$. The non-zero elements of row i in matrix U (without the diagonal elements) are stored between $HA(i,2)$ and $HA(i,3)$. Information concerning the row and column interchanges is stored in $HA(i,7)$ and $HA(i,8)$, $i=1(1)N-1$. Information about the largest number of non-zero elements found in row i / column j during any stage of the elimination is stored (when $IFLAG(4)=1$ only) in $HA(i,9)/HA(j,10)$. The other information stored in array HA is not used in the further computations.

IHA   -   INTEGER. On entry IHA must contain the length of the first dimension of array HA. **Restriction: IHA $\geq$ N. Unchanged on exit.**

AFLAG -   REAL (in the single precision version Y12MCE) or DOUBLE PRECISION (in the double precision version Y12MCF) array of length 8. The content of the array can be described as follows:

       AFLAG(1) -   **Stability factor.** An element can be chosen as pivotal element only if this element is larger (in absolute value) than the absolute value of the largest element in its row divided by AFLAG(1). On entry AFLAG(1) should contain a real number larger than 1.0. If this is not so then the subroutine sets

AFLAG(1) = 1.0005. Recommended values: AFLAG(1) ranging from 4.0 to 16.0. **Unchanged on exit** (when correctly initialized).

AFLAG(2) -   **Drop-tolerance.** An element which in the process of the computations becomes smaller (in absolute value) than the drop-tolerance is removed from array A (and its row and column numbers are removed from arrays RNR and SNR). On entry AFLAG(2) should contain some positive small number or zero. Recommended value AFLAG(2) = 10^{-12} when matrix A is not too badly scaled and the magnitude of the elements is 1, otherwise smaller values of AFLAG(2) should be used. **Unchanged on exit.**

AFLAG(3) -   The subroutine will stop the computation when the growth factor (parameter AFLAG(5), see below) becomes larger than AFLAG(3). On entry AFLAG(3) should contain a large positive number. If AFLAG(3)<10^5 then the subroutine sets AFLAG(3)=10^5. Recommended value AFLAG(3)=10^{16}. **Unchanged on exit** (when correctly initialized).

AFLAG(4) -   The subroutine will stop the computation when the absolute value of a current pivotal element is smaller than AFLAG(4)★AFLAG(6) (parameter AFLAG(6) is described below). On entry AFLAG(4) must contain a small non-negative number. If AFLAG(4)<0 on entry then the subroutine sets AFLAG(4)=-AFLAG(4). Recommended value AFLAG(4)=10^{-12}. **Unchanged on exit** (when correctly initialized).

AFLAG(5) -   **Growth factor. The content of parameter AFLAG(5) is**

modified by subroutine Y12MC. After each stage of the Gaussian elimination subroutine Y12MC sets AFLAG(5) = AFLAG(7)/AFLAG(6) (parameters AFLAG(6) and AFLAG(7) are described below). On exit large values of parameters AFLAG(5) indicate that an appreciable error in the computed solution is possible. In an extreme case , where AFLAG(5)>AFLAG(3) the subroutine will terminate the computations in an attempt to prevent overflow (and IFAIL will be set equal to 4).

AFLAG(6) – On entry AFLAG(6) is equal to the largest element in the coefficient matrix A of the system Ax = b (set by subroutine Y12MB). **Unchanged on exit.**

AFLAG(7) – On exit the largest (in absolute value) element found during any stage of the elimination will be stored in AFLAG(7). **The content of parameter AFLAG(7) is modified by subroutine Y12MC.**

AFLAG(8) – On entry the minimal (in absolute value) pivotal element will be stored in AFLAG(8). Small values of AFLAG(8) indicate numerical singularity of the coefficient matrix A. We advise the user to check this parameter on exit from the calculation very carefully. **The content of parameter AFLAG(8) is modified by the subroutine Y12MC.**

IFLAG – INTEGER array of length 10. **The content of this array is modified by subroutine Y12MC.** The contents of the components of this array can be described as follows.

IFLAG(1) - This parameter is used in connection with the error diagnostics.

The user should set IFLAG(1) $\geq$ 0 before the call of package Y12M (i.e. before the first call of a subroutine of this package). IFLAG(1) is used in the error checks by the subroutines and should not be altered by the user between any two successive calls of subroutines of the package. IFLAG(1) will be equal to -2 on successful exit from subroutine Y12MC. Thus, **the content of IFLAG(1) is modified by subroutine Y12MC.**

IFLAG(2) -　　On entry IFLAG(2) must contain some positive integer smaller than N. We recommend IFLAG(2) $\leq$ 3. If IFLAG(3) = 0 then this parameter is ignored by subroutine Y12MC. If IFLAG((3) $\geq$ 0 then the pivotal search at any stage of the elimination (except possibly some of the last stages) is carried out in the IFLAG(2) rows which have least numbers of non-zero elements. **Unchanged on exit.**

IFLAG(3) -　　On entry IFLAG(3) must contain 0, 1 or 2. For general pivotal search IFLAG(3) should be set equal to 1. If IFLAG(3) = 2 then only diagonal elements of the coefficient matrix A will be selected as pivotal elements. If IFLAG(3) = 0 then the Gaussian elimination will be carried out without any pivoting. IFLAG(3)=0 or IFLAG(3)=2 (i.e. one of the special pivotal strategies is to be applied) should be used very carefully because the error diagnostics algorithm may not trace all errors in this case. For example, if the user attempts to use IFLAG(3)=0 for matrix A which contains zero elements on the main diagonal, then the run will often be stopped because a division by zero occurs. **Unchanged on exit.**

IFLAG(4) -　　On entry IFLAG(4) must contain 0, 1 or 2. IFLAG(4) = 0 is the best choice when (i) only one system is to be solved, (ii) the first system of a sequence of systems with the same matrix ($Ax = b_1$, $Ax = b_2$, ..., $Ax = b_p$) is to be solved, (iii) when any system in a sequence of systems whose matrices are of different structure is to be solved. IFLAG(4) = 1 is the best choice when the first system of a sequence of systems whose matrices are of the same structure is to be solved. In this case IFLAG(4) = 2 is to be used in the solution of any system   after the first one. **Unchanged on exit.**

IFLAG(5) -   On entry IFLAG(5) must contain 1 or 2. If IFLAG(5) = 1 then the non-zero elements of matrix L will be removed. If IFLAG(5) = 2 then the non-zero elements of matrix L will be stored. **Unchanged on exit.**

IFLAG(6) -   On successful exit IFLAG(6) will be equal to the number of "garbage" collections in the row ordered list. If IFLAG(6) is large then it is better to choose a larger value of NN with next calls of subroutine Y12MC with the same or a similar matrix A. This will lead to a reduction in the computing time. **The content of IFLAG(6) is modified by the subroutine Y12MC.**

IFLAG(7) -   On successful exit IFLAG(7) will be equal to the number of "garbage" collections in the column ordered list. If IFLAG(7) is large then it is better to choose a larger value of NN1 in the next calls of subroutine Y12MC with the same or a similar matrix A. This will lead to a reduction in the computing time. **The content of IFLAG(7) is modified by subroutine Y12MC.**

IFLAG(8) -   On successful exit IFLAG(8) will be equal to the maximal number of non-zero elements kept in array A at any stage of the Gaussian elimination. If IFLAG(8) is much smaller then NN (or

NN1) then the length NN (or NN1) can be chosen smaller in next calls of subroutine Y12MC with the same or a similar matrix A. This will lead to a reduction of the storage needed. **The content of IFLAG(8) is modified by subroutine Y12MC.**

IFLAG(9) - **The content of IFLAG(3) is modified by subroutine Y12MC when IFLAG(4) = 1 and ignored otherwise.** The minimal length NN1 such that Y12MC can solve systems whose matrices are of the same structure without "garbage" collections in the column ordered list and "movings" of columns at the end of the column ordered list is stored in IFLAG(9) after the solution of the first system in the sequence (with IFLAG(4) = 1).

IFLAG(10)- **The content of IFLAG(10) is modified by the subroutine Y12MC when IFLAG(4) = 1 and ignored otherwise.** The minimal length NN such that subroutine Y12MC can solve systems whose matrices are of the same structure without "garbage" collections in the row ordered list and "movings" of rows to the end of the row ordered list is stored in IFLAG(10) after the solution of the first system in the sequence (with IFLAG(4) = 1).

IFAIL - **Error diagnostics parameter. The content of parameter IFAIL is modified by subroutine Y12MC.** On exit IFAIL = 0 if the subroutine has not detected any error. Positive values of IFAIL on exit show that some error has been detected by the subroutine. Many of the error diagnostics are common for all subroutines in the package. Therefore the error diagnostics are listed in a separate section, Section 7, of this book. We advise the user to check the value of this parameter on exit.

4.5. Error diagnostics

Error diagnostics are given by positive values of the parameter IFAIL (see above). We advise the user to check carefully the value of this parameter on exit. The error messages are listed in Section 7.

4.6. Auxiliary subroutines

None.

4.7. Timing

The time taken depends on the order of the matrix (parameter N), the number of the non-zero elements in the matrix (parameter Z), the magnitude of the non-zero elements and their distribution in the matrix.

4.8. Storage

There are no internally declared arrays.

4.9. Accuracy

It is difficult to evaluate the accuracy of the computed solution. Large values of parameter AFLAG(5), **the growth factor**, indicate unstable computations during the decomposition stage. Small values of parameter AFLAG(8) can be considered as a signal for numerical singularity. We must emphasize here that normally much more reliable evaluations of the accuracy achieved can be found by the use of iterative refinement, i.e. by the use of subroutine Y12MF. By the use of the latter subroutine the computing time will often be reduced too.

4.10. Some remarks

Remark 1   The values on entry of Y12MC are the same as the values on exit of Y12MB for many parameters. Therefore the content of N, Z, A, SNR, NN, RNR, NN1, columns 1, 3, 4, 6, 7, 8 and 11 of HA, AFLAG(6), IFLAG(1), IFLAG(4) and IFAIL should not be changed between calls of Y12MB and Y12MC.

Remark 2   The call of Y12MC is normally followed by one or several calls of Y12MD. The content of N, A, SNR, NN, B, PIVOT, columns 1, 2, 3, 7 and 8 of HA, IHA, AFLAG, IFLAG(1), IFLAG(3), IFLAG(4) and IFAIL should not be altered between calls of Y12MC and Y12MD.

Remark 3   If IFAIL > 0 on exit from Y12MC there is no justification for calling Y12MD. Therefore the computations should be terminated and the user should investigate (using the error diagnostics given in Section 7) why Y12MC has assigned a positive value to IFAIL.

4.11. Example.

Consider the following system:

$$Ax = b,$$

where

$$A = \begin{matrix} 10 & 0 & 0 & 0 & 0 & 0 \\ 0 & 12 & -3 & -1 & 0 & 0 \\ 0 & 0 & 15 & 0 & 0 & 0 \\ -2 & 0 & 0 & 20 & -2 & 0 \\ -1 & 0 & 0 & -5 & 1 & -1 \\ -1 & -2 & 0 & 0 & 0 & 6 \end{matrix}$$

and

$$b = (10, 11, 45, 33, -22, 31)^{\mathsf{T}}.$$

Input data - Some components of the arrays AFLAG and IFLAG must be initialized before the call of subroutines (see the code). We give below only the data initialized by READ statements.

Output data - The first group of numbers gives information about the positions of the non-zero elements of the matrix U and of matrix L (without the diagonal elements of these matrices), e.g. the numbers on the third line show that at the third position of array A an element with column number two is written and that the element itself is equal to -0.20000. If the second number on a line is zero it means that this position of array A is free. The second group of numbers gives information about the starts and the ends of the rows. If the row start is larger than the row end (i.e. if $HA(I,1) \geq HA(I,3)$) this means that the diagonal element is the only non-zero element in the row (and since this element is stored in array

PIVOT, there is no element of this row in array A). The other information which we give below is clear.

The single precision version Y12MCE is used in the example.

4.11.1. Program

```
      PARAMETER IHA = 10,NN = 200,NN1 = 100
      IMPLICIT REAL(A-B,G,P,T-V),INTEGER(C,F,H-N,R-S,Z)
      REAL A(NN),PIVOT(IHA),B(IHA),AFLAG(8)
      INTEGER SNR(NN),RNR(NN1),HA(IHA,11),IFLAG(10)
      DATA NIN/5/, NOUT/6/
C
C     INITIALIZATION OF THE PARAMETERS.
C
      AFLAG(1) = 16.0
      AFLAG(2) = 1.0E-16
      AFLAG(3) = 1.0E+16
      AFLAG(4) = 1.0E-12
      IFLAG(1) = 0
      IFLAG(2) = 1
      IFLAG(3) = 1
      IFLAG(4) = 1
      IFLAG(5) = 2
      READ (NIN,101)N,Z
101   FORMAT(2I4)
C
C     INITIALIZE THE NON-ZERO ELEMENTS OF MATRIX A
C     IN ARBITRARY ORDER.
C
```

```
        DO 120 K = 1,Z
        READ(NIN,110) RNR(K), SNR(K), A(K)
110     FORMAT(2I4,F12.6)
120     CONTINUE
C
C       INITIALIZE THE COMPONENTS OF THE RIGHT-HAND SIDE VECTOR  B.
C
        READ(NIN,130) (B(K),K = 1,N)
130     FORMAT(6F12.6)
C
C       CALL THE SUBROUTINES  Y12MBE, Y12MCE.
C
        CALL Y12MBE(N,Z,A,SNR,NN,RNR,NN1,HA,IHA,AFLAG,IFLAG,IFAIL)
        IF(IFAIL.NE.0)GO TO 50
        CALL Y12MCE(N,Z,A,SNR,NN,RNR,NN1,PIVOT,B,HA,IHA,
      1 AFLAG,IFLAG,IFAIL)
C
C       PRINT THE RESULTS.
C
50      CONTINUE
        WRITE(NOUT,1)IFAIL
1       FORMAT('1THE ERROR DIAGNOSTIC PARAMETER  IFAIL',
      1 ' IS EQUAL TO',I4)
        IF(IFAIL.GT.0)GO TO 5
        WRITE(NOUT,2)
2       FORMAT('0    I   SNR(I)     A(I)')
        DO 4 I = 1,Z
        WRITE(NOUT,3) I, SNR(I), A(I)
3       FORMAT(' ',2I10,F20.5)
4       CONTINUE
        WRITE(NOUT,21)
```

```
21        FORMAT('0    I    HA(I,1)    HA(I,3)')
          DO 22 I = 1,N
          WRITE(NOUT,23) I, HA(I,1), HA(I,3)
23        FORMAT(3I10)
22        CONTINUE
          WRITE(NOUT,24)
24        FORMAT('0THE PIVOTAL ELEMENTS ARE GIVEN BELOW')
          DO 25 I = 1,N
          WRITE(NOUT,26) I, PIVOT(I)
26        FORMAT(I10,F20.5)
25        CONTINUE
C
C         PRINT THE AUXILIARY INFORMATION ABOUT THE SOLUTION.
C         THIS IS OPTIONALLY.
C
          WRITE(NOUT,11)AFLAG(6)
11        FORMAT('0THE LARGEST ELEMENT IN THE ORIGINAL',
         1 ' MATRIX IS:'F12.5)
          WRITE(NOUT,12)AFLAG(7)
12        FORMAT('0THE LARGEST ELEMENT FOUND IN THE ELIMINATION',F12.5)
          WRITE(NOUT,13)AFLAG(8)
13        FORMAT('0THE MINIMAL(IN ABSOLUTE VALUE)PIVOTAL ELEMENT',F12.5)
          WRITE(NOUT,14)AFLAG(5)
14        FORMAT('0THE GROWTH FACTOR IS:',1PD12.2)
          WRITE(NOUT,15)IFLAG(6)
15        FORMAT('0THE NUMBER OF COLLECTIONS IN THE ROW LIST',I5)
          WRITE(NOUT,16)IFLAG(7)
16        FORMAT('0THE NUMBER OF COLLECTIONS IN THE COLUMN LIST',I5)
          WRITE(NOUT,17)IFLAG(8)
17        FORMAT('0THE LARGEST NUMBER OF ELEMENTS FOUND IN ARRAY A',I9)
5         STOP
```

END

4.11.2. Input

```
6   15
1   1   10.
6   6   6.
6   2   -2.
6   1   -1.
2   2   12.
2   3   -3.
2   4   -1.
4   1   -2.
5   1   -1.
5   6   -1.
5   5   1.
5   4   -5.
4   4   10.
4   5   -1.
3   3   15.
10.        11.        45.        33.        -22.        31.
```

4.11.3. Output

THE ERROR DIAGNOSTIC PARAMETER IFAIL IS EQUAL TO   0

I	SNR(I)	A(I)
1	0	10.00000
2	0	-.20000
3	2	-.20000
4	6	-1.00000
5	0	15.00000
6	0	-.20000
7	1	-.20000
8	6	10.00000
9	1	-.10000
10	3	-1.00000
11	4	-.16667
12	5	-.02778
13	6	.00000
14	1	-.10000
15	5	-2.00000

I	HA(I,1)	HA(I,3)
1	2	1
2	6	5
3	7	8
4	14	15
5	3	4
6	9	13

THE PIVOTAL ELEMENTS ARE GIVEN BELOW

1	10.00000
2	15.00000
3	-1.00000
4	6.00000
5	12.00000
6	4.97222

THE LARGEST ELEMENT IN THE ORIGINAL MATRIX IS: 15.00000

THE LARGEST ELEMENT FOUND IN THE ELIMINATION   15.00000

THE MINIMAL(IN ABSOLUTE VALUE)PIVOTAL ELEMENT   1.00000

THE GROWTH FACTOR IS: 1.00+000

THE NUMBER OF COLLECTIONS IN THE ROW LIST   0

THE NUMBER OF COLLECTIONS IN THE COLUMN LIST   0

THE LARGEST NUMBER OF ELEMENTS FOUND IN ARRAY A   15

5. Documentation of subroutine Y12MD

5.1. Purpose

Y12MD solves sparse systems of linear algebraic equations.

5.2. Calling sequence and declaration of the parameters

The subroutine is written in FORTRAN and it has been extensively tested with the FOR and FTN compilers on the UNIVAC 1100/82 computer at the Regional Computing Centre at the University of Copenhagen (RECKU). Many examples have been run on an IBM 3033 computer at the Northern Europe University Computing Centre (NEUCC) and on a CDC Cyber 173 computer at the Regional Computing Centre at the University of Aarhus (RECAU). Two different versions are available: a single precision version named Y12MDE and a double precision version named Y12MDF. The calls of these two versions and the declarations of the parameters are as follows.

A). Single precision version: Y12MDE

```
      SUBROUTINE Y12MDE(N, A, NN, B, PIVOT, SNR,
     1                  HA, IHA,, IFLAG, IFAIL)
      REAL A(NN), PIVOT(N), B(N)
      INTEGER SNR(NN), HA(IHA,11), IFLAG(10)
      INTEGER N, NN, IHA, IFAIL
```

B). Double precision version: Y12MDF

```
      SUBROUTINE Y12MDF(N, A, NN, B, PIVOT, SNR,
1                       HA, IHA,, IFLAG, IFAIL)
      DOUBLE PRECISION A(NN), PIVOT(N), B(N)
      INTEGER SNR(NN), HA(IHA,11), IFLAG(10)
      INTEGER N, NN, IHA, IFAIL
```

These two versions can be used on many others computers also. However some alterations may be needed and/or may ensure greater efficiency of the performance of the subroutine. For example, it will be much more efficient to declare arrays SNR and (if possible) HA as INTEGER★2 arrays on some IBM installations.

5.3. Method

The LU decomposition (or only the upper triangular matrix U if the vector $c = L^{-1}Pb$ is already computed) is used to obtain an approximation to the solution vector x of the system $Ax = b$. The LU decomposition (or only matrix U) must be available on entry. This means that if a system with a new matrix is solved then the call of subroutine Y12MD must be preceded by a call of Y12MC and the contents of some parameters and arrays should not be altered between the calls of Y12MC and Y12MD. If a system with the same matrix has already been solved, then Y12MD only should be called (but the contents of some parameters and arrays used in the preceding call of Y12MD should not be altered).

5.4. Parameters of the subroutine

N - INTEGER. On entry N must contain the number of equations in the system Ax=b. **Unchanged on exit.**

A - REAL (in the single precision version Y12MDE) or DOUBLE PRECISION (in the double precision version Y12MDF) array of length NN (see below). On entry array A must contain the non-zero elements of the upper triangular matrix U (without the diagonal elements) and sometimes also the non-zero elements of the lower triangular matrix L under the diagonal (these elements are calculated by the subroutine Y12MC, the non-zero elements of L are stored by Y12MC only when this subroutine is called with IFLAG(5) = 2). **The content of the array A is not modified by subroutine Y12MD.**

NN - INTEGER. On entry NN must contain the length of the arrays A and SNR. **Restriction: NN > Z.** Recommended value: $2 \star Z < NN < 3 \star Z$ if the non-zero elements of matrix L will be removed during the decomposition (i.e. if subroutine Y12MC is called with IFLAG(5) = 1) and $3 \star Z < NN < 5 \star Z$ if the non-zero elements of matrix L will be stored during the decomposition (i.e. if the subroutine Y12MC is called with IFLAG(5) = 2). **Unchanged on exit.**

B - REAL (in the single precision version Y12MDE) or DOUBLE PRECISION (in the double precision version Y12MDF) array of length N. If subroutine Y12MC has been called in the solution of system Ax = b (i.e. if IFLAG(5)<3), then on entry array B must contain vector $c = L^{-1}Pb$ (computed by Y12MC). If subroutine Y12MC has not been called in the solution of system Ax = b (i.e a system with the same matrix has been solved before solving Ax = b and the old LU decomposition is used by setting IFLAG(5) = 3, see below) the array B must contain the right hand side vector b on entry. **The content of the array B is modified by the subroutine Y12MD.** On successful exit an approximation to the true solution x will be found in B.

PIVOT -   REAL (in the single precision version Y12MDE) or DOUBLE PRECISION (in the double precision version Y12MDF) array of length N. On entry the diagonal elements of the upper triangular matrix U must be stored in array PIVOT (these elements are calculated and stored in PIVOT by subroutine Y12MC). **Unchanged on exit.**

SNR   -   INTEGER array of length NN (see below). On entry the column numbers of the non-zero elements of the upper triangular matrix U (without the column numbers of the non-zero elements on the diagonal) and sometimes also the column numbers of the non-zero elements of the lower triangular matrix L (without the column numbers of the non-zero elements on the diagonal) must be stored in array SNR. This structure is prepared by Y12MC. **The content of array SNR is not modified by the subroutine Y12MD.**

HA   -   INTEGER two-dimesional array. The length of the first dimension of HA is IHA (see below). The length of the second dimension of HA is 11. The content of columns 4, 5, 6, 9, 10 and 11 is ignored by subroutine Y12MD. The content of columns 1, 2, 3, 7 and 8 is stored by subroutine Y12MC and should not be altered between calls of Y12MC and Y12MD. **Unchanged on exit.**

IHA   -   INTEGER. On entry IHA must contain the length of the first dimension of array HA. **Restriction: IHA $\geq$ N. Unchanged on exit.**

IFLAG -   INTEGER array of length 10. **The contents of all locations in this array except IFLAG(1), IFLAG(3), IFLAG(4) and IFLAG(5) are ignored by subroutine Y12MD.** The user should set IFLAG(1) $\geq$ 0 before the call of package Y12M (i.e. before the first call of a subroutine of this package). IFLAG(1) is used in the error checks by the subroutines and should not be altered by the user between any two successive calls of subroutines of the package. IFLAG(1) is equal to -2 both on entry and on successful exit from subroutine Y12MD. Thus, **the content of IFLAG(1) is not modified by subroutine Y12MD.** The same values of IFLAG(3)

and IFLAG(4) as those in the preceding call of Y12MC should be used. If
subroutine Y12MC has been called in the solution of the system Ax = b, then on
entry IFLAG(5) must have the same value as on entry of Y12MC. If subroutine
Y12MC has not been called in the solution of the system Ax = b (i.e. a system
with the same matrix has been solved before solving Ax = b and the LU
decomposition of matrix A is thus available) then on entry IFLAG(5) must contain
the number 3. The content of array IFLAG is unchanged on exit.

IFAIL - **Error diagnostics parameter. The content of parameter IFAIL is modified by
subroutine Y12MA.** On exit IFAIL = 0 if the subroutine has not detected any
error. Positive values of IFAIL on exit show that some error has been detected by
the subroutine. Many of the error diagnostics are common for all subroutines in
the package. Therefore the error diagnostics are listed in a separate section,
Section 7, of this book. We advise the user to check the value of this parameter
on exit.

5.5. Error diagnostics

Error diagnostics are given by positive values of the parameter IFAIL (see above). We advise
the user to check carefully the value of this parameter on exit. The error messages are listed
in Section 7.

5.6. Auxiliary subroutines

None.

5.7. Timing

The time taken depends on the order of the matrix (parameter N) and the number of the non-zero elements in the LU-decomposition of matrix A.

5.8. Storage

There are no internally declared arrays.

5.9. Accuracy

It is difficult to evaluate the accuracy of the computed solution. Large values of parameter AFLAG(5), **the growth factor**, indicate unstable computations during the decomposition stage. Small values of parameter AFLAG(8), the minimal pivotal element, can be considered as a signal for numerical singularity. We must emphasize here that normally much more reliable evaluations of the accuracy achieved can be found by the use of iterative refinement, i.e. by the use of subroutine Y12MF. By the use of the latter subroutine the computing time will often be reduced too. However, the storage requirements may sometimes be increased.

5.10. Some remarks

Remark 1   The content of N, A, SNR, NN, columns 1, 2, 3, 7 and 8 of HA, IHA, IFLAG(3), IFLAG(4) and IFAIL should not be altered between calls of Y12MC and Y12MD.

Remark 2   If the LU decomposition of matrix A is available (i.e. a system with matrix A has already been solved) then Y12MB and Y12MC should not be called. The user should only assign the new right hand side vector to array B, set IFLAG(5) = 3 and call Y12MD.

5.11. Example.

Consider the following system:

$$Ax = b,$$

where

$$A = \begin{matrix} 10 & 0 & 0 & 0 & 0 & 0 \\ 0 & 12 & -3 & -1 & 0 & 0 \\ 0 & 0 & 15 & 0 & 0 & 0 \\ -2 & 0 & 0 & 20 & -2 & 0 \\ -1 & 0 & 0 & -5 & 1 & -1 \\ -1 & -2 & 0 & 0 & 0 & 6 \end{matrix}$$

and

$$b = (10, 11, 45, 33, -22, 31)^\mathsf{T}.$$

The single precision version Y12MDE is used in the example.

5.11.1. Program

```
      PARAMETER IHA=10,NN=200,NN1=100
      IMPLICIT REAL(A-B,G,P,T-V),INTEGER(C,F,H-N,R-S,Z)
      REAL A(NN),PIVOT(IHA),B(IHA),AFLAG(8)
      INTEGER SNR(NN),RNR(NN1),HA(IHA,11),IFLAG(10)
      DATA NIN/5/, NOUT/6/
C
C     INITIALIZATION OF THE PARAMETERS.
C
      AFLAG(1)=16.
      AFLAG(2)=1.E-16
      AFLAG(3)=1.E+16
      AFLAG(4)=1.E-12
      IFLAG(1)=0
      IFLAG(2)=1
      IFLAG(3)=1
      IFLAG(4)=1
      IFLAG(5)=2
      READ (NIN,101)N,Z
101   FORMAT(2I4)
C
C     INITIALIZE THE NON-ZERO ELEMENTS OF MATRIX A IN ARBITRARY ORDER.
C
      DO 120 K=1,Z
      READ(NIN,110) RNR(K), SNR(K), A(K)
110   FORMAT(2I4,F12.6)
120   CONTINUE
C
C
C     INITIALIZE THE COMPONENTS OF THE RIGHT-HAND SIDE VECTOR B.
```

```
C

        READ(NIN,130) (B(K),K=1,N)
130     FORMAT(6F12.6)
C
C       CALL THE SUBROUTINES Y12MBE, Y12MCE, Y12MDE.
C

        CALL Y12MBE(N,Z,A,SNR,NN,RNR,NN1,HA,IHA,AFLAG,IFLAG,IFAIL)
        IF(IFAIL.NE.0)GO TO 50
        CALL Y12MCE(N,Z,A,SNR,NN,RNR,NN1,PIVOT,B,HA,IHA,AFLAG,IFLAG,IFAIL)
        IF(IFAIL.NE.0)GO TO 50
        CALL Y12MDE(N,A,NN,B,PIVOT,SNR,HA,IHA,IFLAG,IFAIL)
C
C       PRINT THE RESULTS.
C
50      CONTINUE
        WRITE(NOUT,1)IFAIL
1       FORMAT('1THE ERROR DIAGNOSTIC PARAMETER  IFAIL  IS EQUAL TO',I4)
        IF(IFAIL.GT.0)GO TO 5
        WRITE(NOUT,2)
2       FORMAT('0THE SOLUTION VECTOR IS GIVEN BELOW.')
        DO 4 I=1,6
        WRITE(NOUT,3)I,B(I)
3       FORMAT(' ',I10,F20.5)
4       CONTINUE
C
C       PRINT THE AUXILIARY INFORMATION ABOUT THE SOLUTION.
C       THIS IS OPTIONALLY.
C
        WRITE(NOUT,11)AFLAG(6)
11      FORMAT('0THE LARGEST ELEMENT IN THE ORIGINAL MATRIX IS:'F12.5)
        WRITE(NOUT,12)AFLAG(7)
```

```
12        FORMAT('0THE LARGEST ELEMENT FOUND IN THE ELIMINATION',F12.5)
          WRITE(NOUT,13)AFLAG(8)
13        FORMAT('0THE MINIMAL(IN ABSOLUTE VALUE)PIVOTAL ELEMENT',F12.5)
          WRITE(NOUT,14)AFLAG(5)
14        FORMAT('0THE GROWTH FACTOR IS:',1PE12.2)
          WRITE(NOUT,15)IFLAG(6)
15        FORMAT('0THE NUMBER OF COLLECTIONS IN THE ROW LIST',I5)
          WRITE(NOUT,16)IFLAG(7)
16        FORMAT('0THE NUMBER OF COLLECTIONS IN THE COLUMN LIST',I5)
          WRITE(NOUT,17)IFLAG(8)
17        FORMAT('0THE LARGEST NUMBER OF ELEMENTS FOUND IN ARRAY A',I9)
5         STOP
          END
```

5.11.2. Input

```
6  15
1   1  10.
6   6   6.
6   2  -2.
6   1  -1.
2   2  12.
2   3  -3.
2   4  -1.
4   1  -2.
5   1  -1.
5   6  -1.
5   5   1.
5   4  -5.
4   4  10.
4   5  -1.
3   3  15.
   10.      11.      45.      33.     -22.      31.
```

5.11.3. Output

THE ERROR DIAGNOSTIC PARAMETER IFAIL IS EQUAL TO   0
THE SOLUTION VECTOR IS GIVEN BELOW.

1	1.00000
2	2.00000
3	3.00000
4	4.00000
5	5.00000
6	6.00000

THE LARGEST ELEMENT IN THE ORIGINAL MATRIX IS: 15.00000
THE LARGEST ELEMENT FOUND IN THE ELIMINATION   15.00000
THE MINIMAL(IN ABSOLUTE VALUE)PIVOTAL ELEMENT   1.00000
THE GROWTH FACTOR IS: 1.00+000
THE NUMBER OF COLLECTIONS IN THE ROW LIST   0
THE NUMBER OF COLLECTIONS IN THE COLUMN LIST   0
THE LARGEST NUMBER OF ELEMENTS FOUND IN ARRAY A   15

6. Documentation of subroutine Y12MF

6.1. Purpose

Large and sparse systems of linear algebraic equations with real coefficients are solved by the use of Gaussian elimination and sparse matrix technique. Iterative refinement is applied in order to improve the accuracy.

6.2. Calling sequence and declaration of the parameters

The subroutine is written in FORTRAN and has been extensively tested with the FOR and FTN compilers on the UNIVAC 1100/82 computer at the Regional Computing Centre at the University of Copenhagen (RECKU). Many examples have been run on an IBM 3033 computer at the Northern Europe University Computing Centre (NEUCC) and on a CDC Cyber 173 computer at the Regional Computing Centre at the University of Aarhus (RECAU). Only a single precision version, Y12MFE, of this subroutine is available. The call of the subroutine and the declarations of the parameters are as follows.

```
       SUBROUTINE Y12MFE(N, A, SNR, NN, RNR, NN1, A1, SN, NZ,
     1                   HA, IHA, B, B1, X, Y, AFLAG, IFLAG, IFAIL)
       REAL A(NN), B(N), B1(N), X(N), Y(N), A1(NZ), AFLAG(11)
       INTEGER SNR(NN), RNR(NN1), HA(IHA,13), SN(NZ), IFLAG(12)
       INTEGER N, NN, NN1, NZ, IHA, IFAIL
```

This subroutine can be used on many other computers also. However, some alterations may be needed and/or may ensure greater efficiency of the performance of the subroutine. For example, it will be much more efficient to declare arrays SNR, RNR, SN and (if possible) HA as INTEGER★2 arrays on some IBM instalations. Note too that the subroutine accumulates some inner products in double precision and then rounds them to single precision. Therefore if the use of double precision is very expensive (in comparison with the use of single precision) on the computer at the user's disposal then it will not be efficient to use subroutine Y12MF. The single precision versions of the other subroutines should be used in this situation.

6.3. Method

Consider the system $Ax = b$, where matrix A is sparse. The non-zero elements of matrix A are ordered by rows (subroutine Y12MB is called to perform this operation) and then factorized into two triangular matrices, an upper triangular matrix U and a unit lower triangular matrix L (subroutine Y12MC is called to calculate the non-zero elements of U amd L). The factorized system is solved (subroutine Y12MD is called to calculate an approximation x_1 to the solution vector x) and an attempt to improve the first solution by iterative refinement is carried out in the following way:

$$r_{k-1} = b - Ax_{k-1}$$

$$d_{k-1} = QU^{-1}L^{-1}Pr_{k-1}$$

$$x_k = x_{k-1} + d_{k-1}$$

$$\text{where } k = 2, 3, ..., p.$$

Normally, this process will improve the accuracy of the first solution x_1. If the process is not convergent or the convegence is very slow, then information about the trouble can be obtained by checking the parameters AFLAG(10) (the max-norm of the last residual vector

r_{p-1} computed by Y12MF) and AFLAG(9) (the max-norm of the last correction vector, d_{p-1}, computed by Y12MF). We advise the user to check carefully these parameters on exit. Large values of these parameters show that the computed solution is not very accurate.

6.4. Parameters of the subroutine

N   -   INTEGER. On entry N must contain the number of equations in the system Ax=b. **Unchanged on exit.**

A   -   REAL array of length NN (see below). If the coefficient matrix must be factorized (in this case IFLAG(5) = 2), then on entry the first NZ locations of array A must contain the non-zero elements of matrix A, the order of the elements can be arbitrary. If the factorization of matrix A is available (i.e. a system with the same matrix has been solved in a previous call of Y12MF) then on entry array A must contain the non-zero elements of U and L (IFLAG(5) = 3 should also be assigned in this case). **The content of array A is modified by Y12MF when the coefficient matrix has to be factorized. Unchanged on exit when an old LU factorization is used.** The content of array A should not be altered between successive calls of Y12MF for the solution of systems with the same matrices.

SNR   -   INTEGER array of the length NN (see below). If the coefficient matrix A must be factorized then on entry the first NZ positions of array A must contain the column numbers of the non-zero elements of matrix A (ordered as the non-zero elements in array A). **The content of array SNR is modified by the subroutine Y12MF in this case.** If the LU factorization of matrix A is available then on entry array SNR must contain the column numbers of the non-zero elements of U and L. **The content of array SNR is unchanged on exit in this case.** The content of array SNR should not be altered between successive calls of Y12MF for the solution of systems with the same matrices.

NN   - INTEGER. On entry NN must contain the length of array A and SNR. **Restriction: NN** $\geq$ **2★Z.** Recommended value: $2 \star Z \leq NN \leq 3 \star Z$ **Unchanged on exit.**

RNR   - INTEGER array of length NN1 (see below). If the coefficient matrix A must be factorized, then on entry the first NZ positions of array RNR must contain the row numbers of the non-zero elements of matrix A (ordered as the non-zero elements in array A). **The content of array RNR is modified by the subroutine in this case.** If the LU factorization of matrix A is available then the content of array RNR is ignored by the subroutine Y12MF.

NN1   - INTEGER. On entry NN1 must contain the length of the array RNR. **Restriction: NN1** $\geq$ **Z.** Recommended value: $1.5 \star Z \leq NN1 \leq 2 \star Z$. **Unchanged on exit.**

A1   - REAL array of length NZ (see below). **The content of array A1 is modified by subroutine Y12MF when the LU factorization of the coefficient matrix A is not available.** Subroutine Y12MF copies the first NZ locations of array A into array A1 in this case (after the internal call of Y12MB; this means that array A1 contains the non-zero elements of the coefficient matrix A ordered by rows on exit). If the LU factorization of the coefficient matrix A is available then **the content of array A1 is unchanged on exit.** The content of array A1 should not be altered between successve calls of Y12MF in the solution of systems with the same matrices.

SN   - INTEGER array of length NZ (see below). **The content of array SN is modified by subroutine Y12MF when the LU factorization of the coefficient matrix A is not available.** Subroutine Y12MF copies the first NZ locations of array SNR into array SN in this case (after the internal call of Y12MB; this means that array SN contains the column numbers of the non-zero elements ordered by rows on exit). If the LU factorization is available then **the content of array SN is unchanged on exit.** The content of array SN should not be altered between successive calls of Y12MF in the solution of systems with the same matrices.

NZ   -   INTEGER. On entry NZ must contain the number of non-zero elements in the
            coefficient matrix A of the system Ax = b. **Unchanged on exit.**

HA   -   INTEGER two-dimensional array. The length of the first dimension of HA is IHA
            (see below). The length of the second dimension of HA is 13. **If a new**
            decomposition should be calculated, then subroutine Y12MF modifies the
            contents of HA. The contents on exit of some of the columns of array HA are
            described in the documentation of subroutine Y12MC. The last two columns
            (12'th and 13'th) contain on exit information about the row starts and the row
            ends in the original matrix (after the non-zero elements of this matrix have been
            ordered by rows); this means that the non-zero elements in row i of the
            coefficient matrix A are located between positions HA(i,12) and HA(i,13) in array
            A1 (the column numbers of the non-zero elements in row i are also located
            between positions HA(i,12) and HA(i,13) in array SN). If the matrix of the system
            has the same structure as the matrix of a system which is already solved, then
            subroutine Y12MF will use the information stored in columns 7, 8, 9 and 10 of
            HA during the previous call (therefore this information should not be altered). **If**
            the LU decomposition of the coefficient matrix is available, then the
            contents of array HA are unchanged on exit. The contents of columns 1, 2, 3,
            7, 8, 12 and 13 of array HA should not be altered between successive calls of
            Y12MF in the solution of systems with the same matrix.

IHA   -   INTEGER. On entry IHA must contain the length of the first dimension of array
            HA. **Restriction: IHA** $\geq$ **N. Unchanged on exit.**

B   -   REAL array of length N. On entry the right-hand side vector b of the system
            Ax=b must be stored in array B. **The content of array B is modified by**
            subroutine Y12MF. On successful exit the components of the last correction
            vector d_{p-1} are stored in array B.

B1 -   REAL array of length N. **The content of array B1 is modified by subroutine**

Y12MF. The right hand side vector of the system Ax = b is stored in array B1 on successful exit.

X - REAL array of length N. **The content of array X is modified by the subroutine Y12MF.** On successful exit the corrected solution vector is stored in array X.

Y - REAL array of length N. **The content of array Y is modified by subroutine Y12MF.** On successful exit array Y will contain the pivotal elements (the diagonal elements of matrix U). This means that a small element (or small elements) in array Y on exit may indicate numerical singularity of the coefficient matrix A. Note that the smallest in absolute value element in array Y is also stored in AFLAG(8), see below.

AFLAG - REAL array of length 11. The content of the array can be described as follows:

AFLAG(1) - **Stability factor.** An element can be chosen as pivotal element only if this element is larger (in absolute value) than the absolute value of the largest element in its row divided by AFLAG(1). On entry AFLAG(1) should contain a real number larger than 1.0. If this is not so then the subroutine sets AFLAG(1) = 1.0005. Recommended values of AFLAG(1) ranging from 4.0 to 16.0. **Unchanged on exit (when correctly initialized).**

AFLAG(2) - **Drop-tolerance.** An element which in the process of the computations becomes smaller(in absolute value) than the drop-tolerance is removed from array A (and its row and column numbers are removed from arrays RNR and SNR). Recommended value: on entry AFLAG(2) should be in the interval $[a \star 10^{-4}, a \star 10^{-1}]$ where a is the magnitude of the elements. If the magnitude a of the non-zero elements is not known, then the user should intialize the required drop-tolerance multiplied by -1. Let a_i be the maximal element (in absolute value) in row i (i=1(1)N). Let a be the minimal number among all a_i. If a negative value of AFLAG(2) is assigned, the subroutine computes   a   and uses a drop-tolerance equal to   -AFLAG(2)$\star$a. In the above considerations it is assumed that the coefficient matrix A is not too badly scaled. If this is not so then perform row scaling of the coefficient matrix so that the largest elements in the rows are of magnitude 1 and choose AFLAG(2) in the interval $[10^{-5},10^{-3}]$. **Unchanged on exit.**

AFLAG(3) -   The subroutine will stop the computation when the growth factor (parameter AFLAG(5), see below) becomes larger than AFLAG(3). On entry AFLAG(3) should contain a large positive number. If   AFLAG(3)$<10^5$ then the subroutine sets AFLAG(3)=10^5. Recommended value   AFLAG(3)=10^{16}. **Unchanged on exit** (when correctly initialized).

AFLAG(4) -   The subroutine will stop the computation when the absolute value of a current pivotal element is smaller than AFLAG(4)$\star$AFLAG(6) (parameter AFLAG(6) is described below). On entry AFLAG(4) must contain a small non-negative number. If   AFLAG(4)<0 then the subroutine sets AFLAG(4)=-AFLAG(4).   Recommended   value AFLAG(4)=10^{-12}. **Unchanged on exit** (when correctly intialized).

AFLAG(5) -   **Growth factor.** The content of parameter AFLAG(5) is modified by subroutine Y12MF. After each stage of the Gaussian elimination subroutine Y12MF sets AFLAG(5) =

AFLAG(7)/AFLAG(6) (parameters AFLAG(6) and AFLAG(7) are described below). On exit large values of parameters AFLAG(5) indicate that an appreciable error in the computed solution is possible. In an extreme case, where AFLAG(5)>AFLAG(3), the subroutine will terminate the computations in an attempt to prevent overflow.

AFLAG(6) -   On exit AFLAG(6) is equal to the largest element in the coefficient matrix A of the system Ax=b (set by subroutine Y12MB). **Unchanged on exit.**

AFLAG(7) -   On exit the largest (in absolute value) element found during any stage of the elimination will be stored in AFLAG(7). **The content of parameter AFLAG(7) is modified by subroutine Y12MF.**

AFLAG(8) -   On exit the minimal (in absolute value) pivotal element will be stored in AFLAG(8). Small values of AFLAG(8) indicate numerical singularity of the coefficient matrix A. We advise the user to check this parameter on exit from the calculation very carefully. **The content of parameter AFLAG(8) is modified by the subroutine Y12MF.**

AFLAG(9) -   **The content of AFLAG(9) is modified by the subroutine Y12MF.** The max-norm of the last correction vector d_{p-1} will be stored in AFLAG(9) on successful exit. We advise the user to check carefully this parameter.

AFLAG(10) -   **The content of AFLAG(10) is modified by the subroutine Y12MF.** The max-norm of the last residual vector r_{p-1} will be stored in AFLAG(10) on successful exit. We advise the user to check carefully this parameter.

AFLAG(11) -   **The content of AFLAG(11) is modified by the subroutine Y12MF.** On exit AFLAG(11) will contain the max-norm of the corrected solution vector. If the value of AFLAG(11) is not to close to zero then AFLAG(9)/AFLAG(11) will normally give an excellent evaluation of the relative error in the solution vector.

IFLAG -   INTEGER array of length 12. The content of this array can be described as follows:

IFLAG(1) -   This parameter is used as a work space by subroutine Y12MF.

IFLAG(2) -   On entry IFLAG(2) must contain some positive integer smaller than N. We recommend IFLAG(2) $\leq$ 3. If IFLAG(3) $=$ 0 then this parameter is ignored by subroutine Y12MF. If IFLAG((2) $\geq$ 0 then the pivotal search at any stage of the elimination (except possibly some of the last stages) is carried out in the IFLAG(2) rows which have least number of non-zero elements. **Unchanged on exit.**

IFLAG(3) -   On entry IFLAG(3) must contain 0, 1 or 2. For general pivotal search IFLAG(3) should be set equal to 1. If IFLAG(3) $=$ 2 then only diagonal elements of the coefficient matrix A can be selected as pivotal elements. If IFLAG(3) $=$ 0 then the Gaussian elimination will be carried out without any pivoting. IFLAG(3)$=$0 or IFLAG(3)$=$2 (i.e. one of the special pivotal strategies is to be applied) should be used very carefully because the error diagnostics algorithm may not trace all errors in this case. **Unchanged on exit.**

IFLAG(4) -   On entry IFLAG(4) must contain 0, 1 or 2. IFLAG(4) = 0 is the best choice when (i) only one system is to be solved, (ii) the first system of a sequence of systems with the same matrix ($Ax = b_1$, $Ax = b_2$,......$Ax = b_p$) is to be solved, (iii) when any system in a sequence of systems whose matrices are of different structure is to be solved. IFLAG(4) = 1 is the best choice when the first system of a sequence of systems whose matrices are of the same structure is to be solved. In this case IFLAG(4) = 2 can be used in the solution of any system   after the first one. **Unchanged on exit.**

IFLAG(5) -   If the LU factorization of the coefficient matrix is not available, then IFLAG(5) must be set to 2 on entry. If the LU factorization of the coefficient matrix is available, then IFLAG(5) must be set to 3 on entry. **Unchanged on exit.**

IFLAG(6) -   On successful exit IFLAG(6) will be equal to the number of "garbage" collections in the row ordered list. If IFLAG(6) is large then it is better to choose a larger value of NN with next calls of subroutine Y12MF with the same or similar matrix A. This will lead to a reduction in the computing time. **The content of IFLAG(6) is modified by the subroutine Y12MF.**

IFLAG(7) -   On successful exit IFLAG(7) will be equal to the number of "garbage" collections in the column ordered list. If IFLAG(7) is large then it is better to choose a larger value of NN1 in the next calls of subroutine Y12MF with the same or similar matrix A. This will lead to a reduction of the computing time. **The content of IFLAG(7) is modified by subroutine Y12MF.**

IFLAG(8) -   On successful exit IFLAG(8) will be equal to the maximal

IFLAG(8) -   number of non-zero elements kept in array A at any stage of the Gaussian elimination. If IFLAG(8) is much smaller than NN (or NN1) then the length NN (or NN1) can be chosen smaller in next calls of subroutine Y12MF with the same or similar matrix A. This will lead to a reduction of the storage needed. **The content of IFLAG(8) is modified by subroutine Y12MF.**

IFLAG(9) -   The minimal length NN1 such that Y12MF can solve systems whose matrices are of the same structure without "garbage" collections in the column ordered list and "movings" of columns at the end of the column ordered list is stored in IFLAG(9) after the solution of the first system in the sequence (with IFLAG(4)=1). **The content of IFLAG(9) is modified by subroutine Y12MF when IFLAG(4) = 1 and ignored otherwise.**

IFLAG(10) -   The minimal length NN such that subroutine Y12MF can solve systems whose matrices are of the same structure without "garbage" collections in the row ordered list and "movings" of rows to the end of the row ordered list is stored in IFLAG(10) after the solution of the first system in the sequence (with IFLAG(4) = 1). **The content of IFLAG(10) is modified by the subroutine Y12MF when IFLAG(4) = 1 and ignored otherwise.**

IFLAG(11) -   The maximum allowed number of iterations must be contained in IFLAG(11) on entry. **Restriction: IFLAG(11)>1.** Recommended value : IFLAG(11)< 33. **Unchanged on exit.**

IFLAG(12) -   **The content of IFLAG(12) is modified by the subroutine Y12MF.** On exit the number of iterations actually performed will be stored in IFLAG(12).

IFAIL -   **Error diagnostic parameter. The content of parameter IFAIL is modified by**

            subroutine Y12MF. On exit IFAIL = 0 if the subroutine has not detected any

            error. Positive values of IFAIL on exit show that some error has been detected by

            the subroutine. Many of the error diagnostics are common for all subroutines in

            the package. Therefore the error diagnostics are listed in a separate section,

            Section 7, of this book. We advise the user to check the value of this parameter

            on exit.

6.5. Error diagnostics

Error diagnostics are given by positive values of parameter IFAIL (see above). We advise the
user to check carefully the value of this parameter on exit. The error messages are listed in
Section 7.

6.6. Auxiliary subroutines

Y12MF calls three other subroutines: Y12MB, Y12MC and Y12MD.

6.7. Timing

The time taken depends on the order of the matrix (parameter N), the number of the non-zero
elements in the matrix (parameter Z), the magnitude of the non-zero elements and their
distribution in the matrix.

6.8. Storage

There are no internally declared arrays.

6.9. Accuracy

Normally the accuracy achieved can be estimated very well by means of AFLAG(9) and AFLAG(10). Further information about the accuracy can sometimes be obtained by inspection of the growth factor AFLAG(5), and the smallest pivotal element (in absolute value) stored in AFLAG(8).

6.10. Some remarks

Remark 1   Following the recommendations given in Section 1.3.6 the user can write a subroutine (similar to the subroutine Y12MA) where the recommended values of the parameters AFLAG(1) to AFLAG(4), IFLAG(2) to IFLAG(5) and IFLAG(11) are initialized. If this is done then only N, NN, NN1, IHA, A, SNR, RNR and B should be assigned before the call of this new subroutine.

Remark 2   The information stored in some of the arrays should not be altered between successive calls of subroutine Y12MF (see more details in Section 1.3.6).

6.11. Example

Consider the following system:

$$Ax = b,$$

where

$$
A = \begin{matrix}
10 & 0 & 0 & 0 & 0 & 0 \\
0 & 12 & -3 & -1 & 0 & 0 \\
0 & 0 & 15 & 0 & 0 & 0 \\
-2 & 0 & 0 & 20 & -2 & 0 \\
-1 & 0 & 0 & -5 & 1 & -1 \\
-1 & -2 & 0 & 0 & 0 & 6
\end{matrix}
$$

and

$$b = (10, 11, 45, 33, -22, 31)^T.$$

6.11.1. Program

```
PARAMETER IHA=10,NN=200,NN1=100,IHA1=60
IMPLICIT REAL(A-B,G,P,T-V),INTEGER(C,F,H-N,R-S,Z)
REAL A(NN),Y(IHA),B(IHA),AFLAG(11),B1(IHA),X(IHA),A1(IHA1)
INTEGER SNR(NN),RNR(NN1),HA(IHA,13),IFLAG(12),SN(IHA1)
DATA NIN/5/, NOUT/6/
C
C      INITIALIZATION OF THE PARAMETERS.
```

```
C

      AFLAG(1)=128.0

      AFLAG(2)=1.E-3

      AFLAG(3)=1.E+16

      AFLAG(4)=1.E-12

      IFLAG(2)=2

      IFLAG(3)=1

      IFLAG(4)=1

      IFLAG(5)=2

      IFLAG(11)=25

      READ (NIN,101)N,Z

101   FORMAT(2I4)

C

C     INITIALIZE THE NON-ZERO ELEMENTS OF MATRIX  A  IN ARBITRARY ORDER.

C

      DO 120 K=1,Z

      READ(NIN,110) RNR(K), SNR(K), A(K)

110   FORMAT(2I4,F12.6)

120   CONTINUE

C

C

C     INITIALIZE THE COMPONENTS OF THE RIGHT-HAND SIDE VECTOR  B.

C

      READ(NIN,130) (B(K),K=1,N)

130   FORMAT(6F12.6)

C

C     CALL THE SUBROUTINE  Y12MFE.

C

      CALL Y12MFE(N,A,SNR,NN,RNR,NN1,A1,SN,Z,HA,IHA,B,B1,X,Y,

     1 AFLAG,IFLAG,IFAIL)

C
```

```
C       PRINT THE RESULTS.
C
        WRITE(NOUT,1)IFAIL
1       FORMAT('1THE ERROR DIAGNOSTIC PARAMETER  IFAIL  IS EQUAL TO',I4)
        IF(IFAIL.GT.0)GO TO 5
        WRITE(NOUT,2)
2       FORMAT('0THE SOLUTION VECTOR IS GIVEN BELOW.')
        DO 4 I=1,6
        WRITE(NOUT,3)I,X(I)
3       FORMAT(' ',I10,F20.5)
4       CONTINUE
C
C       PRINT THE AUXILIARY INFORMATION ABOUT THE SOLUTION.
C       THIS IS OPTIONALLY.
C
        WRITE(NOUT,11)AFLAG(6)
11      FORMAT('0THE LARGEST ELEMENT IN THE ORIGINAL MATRIX IS:'F12.5)
        WRITE(NOUT,12)AFLAG(7)
12      FORMAT('0THE LARGEST ELEMENT FOUND IN THE ELIMINATION',F12.5)
        WRITE(NOUT,13)AFLAG(8)
13      FORMAT('0THE MINIMAL(IN ABSOLUTE VALUE)PIVOTAL ELEMENT',F12.5)
        WRITE(NOUT,14)AFLAG(5)
14      FORMAT('0THE GROWTH FACTOR IS:',1PE12.2)
        WRITE(NOUT,15)IFLAG(6)
15      FORMAT('0THE NUMBER OF COLLECTIONS IN THE ROW LIST',I5)
        WRITE(NOUT,16)IFLAG(7)
16      FORMAT('0THE NUMBER OF COLLECTIONS IN THE COLUMN LIST',I5)
        WRITE(NOUT,17)IFLAG(8)
17      FORMAT('0THE LARGEST NUMBER OF ELEMENTS FOUND IN ARRAY A',I9)
        WRITE(NOUT,18) IFLAG(12)
18      FORMAT('0THE NUMBER OF ITERATIONS IS:,'I5)
```

```
       WRITE(NOUT,19) AFLAG(9)
19     FORMAT('0ESTIMATION OF THE ABSOLUTE ERROR IS:',1PE10.2)
5      CONTINUE
       STOP
       END
```

6.11.2. Input

```
6  15
1   1  10.
6   6   6.
6   2  -2.
6   1  -1.
2   2  12.
2   3  -3.
2   4  -1.
4   1  -2.
5   1  -1.
5   6  -1.
5   5   1.
5   4  -5.
4   4  10.
4   5  -1.
3   3  15.
10.      11.      45.      33.      -22.      31.
```

6.11.3. Output

```
THE ERROR DIAGNOSTIC PARAMETER  IFAIL  IS EQUAL TO    0
THE SOLUTION VECTOR IS GIVEN BELOW.
        1            1.00000
        2            2.00000
        3            3.00000
        4            4.00000
        5            5.00000
        6            6.00000
THE LARGEST ELEMENT IN THE ORIGINAL MATRIX IS:    15.00000
THE LARGEST ELEMENT FOUND IN THE ELIMINATION    15.00000
THE MINIMAL(IN ABSOLUTE VALUE)PIVOTAL ELEMENT      .49722
THE GROWTH FACTOR IS:    1.00+000
THE NUMBER OF COLLECTIONS IN THE ROW LIST    0
THE NUMBER OF COLLECTIONS IN THE COLUMN LIST    0
THE LARGEST NUMBER OF ELEMENTS FOUND IN ARRAY A      15
THE NUMBER OF ITERATIONS IS:,    2
ESTIMATION OF THE ABSOLUTE ERROR IS: 2.38-006
```

7. Error diagnostics

IFAIL is the error diagnostics parameter. On exit from each subroutine IFAIL = 0 if no error has been detected. IFAIL > 0 indicates that some error has been detected and the computations have been terminated immediately after the detection of the error. The errors corresponding to the different positive values of IFAIL are listed below.

IFAIL = 1   The coefficient matrix A is not factorized, i.e. the call of subroutine Y12MD was not preceded by a call of Y12MC during the solution of Ax=b or during the solution of the first system in a sequence ($Ax_1 = b_1$, $Ax_2 = b_2,......,Ax_p = b_p$) of systems with the same coefficient matrix. This will work in all cases only if the user sets IFLAG(1) $\geq$ 0 before the call of package Y12M (i.e. before the first call of a subroutine of this package).

IFAIL = 2   The coefficient matrix A is not ordered, i.e. the call of subroutine Y12MC was not preceded by a call of Y12MB. This will work in all cases only if the user sets IFLAG(1) $\geq$ 0 before the call of package Y12M (i.e. before the first call of a subroutine of this package).

IFAIL = 3   A pivotal element $|a_{ii}^{(p)}| <$ AFLAG(4) $\star$ AFLAG(6) is selected. When AFLAG(4) is sufficiently small this is an indication that the coefficient matrix is numerically singular.

IFAIL = 4   AFLAG(5), the growth factor, is larger than AFLAG(3). When AFLAG(3) is sufficiently large this indicates that the elements of the coefficient matrix A grow so quickly during the factorization that the continuation of the computation is not justified. The choice of a smaller stability factor, AFLAG(1), may give better results in this case.

IFAIL = 5   The length NN of arrays A and SNR is not sufficient. Larger values of NN (and possibly of NN1) should be used.

IFAIL = 6   The length NN1 of array RNR is not sufficient. Larger values of NN1 (and possibly of NN) should be used.

IFAIL = 7   A row without non-zero elements in its active part is found during the decomposition. If the drop-tolerance, AFLAG(2), is sufficiently small, then IFAIL = 7 indicates that the matrix is numerically singular. If a large value of the drop-tolerance AFLAG(2) is used and if IFAIL = 7 on exit, this is not certain. A run with a smaller value of AFLAG(2) and/or a careful check of the parameters AFLAG(8) and AFLAG(5) is recommended in the latter case.

IFAIL = 8   A column without non-zero elements in its active part is found during the decomposition. If the drop-tolerance, AFLAG(2), is sufficiently small, then IFAIL = 8 indicates that the matrix is numerically singular. If a large value of the drop-tolerance AFLAG(2) is used and if IFAIL = 8 on exit, this is not certain. A run with a smaller value of AFLAG(2) and/or a careful check of the parameters AFLAG(8) and AFLAG(5) is recommended in the latter case.

IFAIL = 9   A pivotal element is missing. This may occur if AFLAG(2) > 0 and IFLAG(4) = 2 (i.e. some system after the first one in a sequence of systems with the same structure is solved using a positive value for the drop-tolerance). The value of the drop-tolerance AFLAG(2), should be decreased and the coefficient matrix of the system refactorized. This error may also occur when one of the special pivotal strategies (IFLAG(3)=0 or

IFAIL = 9   IFLAG(3)=2) is used and the matrix is not suitable for such a strategy.

IFAIL = 10   Subroutine Y12MF is called with IFLAG(5) = 1 (i.e. with a request to
                 remove the non-zero elements of the lower triangular matrix L).
                 IFLAG(5)=2 must be initialized instead of IFLAG(5)=1.

IFAIL = 11   The coefficient matrix A contains at least two elements in the same
                 position (i,j). The input data should be examined carefully in this case.

IFAIL = 12   The number of equations in the system Ax=b is smaller than 2 (i.e.
                 N<2). The value of N should be checked.

IFAIL = 13   The number of non-zero elements of the coefficient matrix is non-positive
                 (i.e. $Z \leq 0$). The value of the parameter Z (renamed NZ in Y12MF) should
                 be checked.

IFAIL = 14   The number of non-zero elements in the coefficient matrix is smaller than
                 the number of equations (i.e. Z < N). If there is no mistake (i.e. if
                 parameter Z, renamed NZ in Y12MF, is correctly assigned on entry) then
                 the coefficient matrix is structurally singular in this case.

IFAIL = 15   The length IHA of the first dimension of array HA is smaller than N.
                 $IHA \geq N$ should be assigned.

IFAIL = 16   The value of parameter IFLAG(4) is not assigned correctly. IFLAG(4)
                 should be equal to 0, 1 or 2. See more details in the description of this
                 parameter.

IFAIL = 17   A row without non-zero elements has been found in the coefficient matrix
                 A of the system before the Gaussian elimination is initiated. Matrix A is
                 structurally singular.

IFAIL = 18   A column without non-zero elements has been found in the coefficient matrix A of the system before the Gaussian elimination is initiated. Matrix A is structurally singular.

IFAIL = 19   Parameter IFLAG(2) is smaller than 1. The value of IFLAG(2) should be a positive integer (IFLAG(2) = 3 is recommended).

IFAIL = 20   Parameter IFLAG(3) is out of range. IFLAG(3) should be equal to 0, 1 or 2.

IFAIL = 21   Parameter IFLAG(5) is out of range. IFLAG(5) should be equal to 1, 2 or 3 (but when IFLAG(5) = 3 Y12MB and Y12MC should not be called; see also the message for IFAIL = 22 below).

IFAIL = 22   Either subroutine Y12MB or subroutine Y12MC is called with IFLAG(5) = 3. Each of these subroutines should be called with IFLAG(5) equal to 1 or 2.

IFAIL = 23   The number of allowed iterations (parameter IFLAG(11) when Y12MF is used) is smaller than 2. IFLAG(11) $\geq$ 2 should be assigned.

IFAIL = 24   At least one element whose column number is either larger than N or smaller than 1 is found.

IFAIL = 25   At least one element whose row number is either larger than N or smaller than 1 is found.

8. Portability of the package

Only standard FORTRAN statements are used in the package. No machine dependent constants are needed. Therefore the subroutines of the package should work well on many large computers. Runs on three different computers have been carried out in order to test the portability of the subroutines of Y12M. Several different compilers have been used on each of these computers. No changes were made in order to perform these runs. Some of the results are summarized below.

8.1. Runs on UNIVAC 1100 series

For the users of RECKU we have performed a comparison of some different compilers in order to select the best for the subroutines of our package. The results of this comparison are given in Table 7 and Table 8. It must be mentioned here that the computations were carried out in July 1978.

COMPILER	FOR		FTN		FTN,V	
N	CLASS D	CLASS E	CLASS D	CLASS E	CLASS D	CLASS E
250	5.20	3.35	6.84	4.44	6.16	3.94
300	6.43	4.35	8.25	5.91	7.41	5.18
350	7.24	5.57	9.35	7.46	8.37	6.63
400	7.67	6.91	10.04	9.28	8.97	8.26
450	8.99	8.27	11.76	11.08	10.54	9.87
500	9.40	10.03	12.33	13.20	10.99	11.85
550	10.12	11.40	12.62	15.03	11.83	13.41
600	11.00	12.95	14.49	17.09	12.92	15.26
Total	66.05	62.83	85.68	83.49	77.19	74.40

Table 7

The computing time when different compilers are used (the time given with each N is the sum of the computing times during the solution of six systems for C = 4, 44, 84, 124, 164 and 204). IR with a stability factor u = 128 and a drop-tolerance T = 10^{-3} has been used. Two rows with minimal numbers of non-zero elements are investigated at each stage of the Gaussian elimination.

COMPILER	TIME	PRICE
FOR	128.88	46.85
FTN	169.17	59.25
FTN,V	151.59	52.63

Table 8

The sum of the computing times (in secs.) and the price (for the total run in Danish kroner) for all 96 systems are given in this table.

For the users at RECKU a version of the iterative refinement subroutine, that exploits the Univac facility: "multibanking", has been developed. In this version two banks are used for the arrays. In the first bank all arrays which are used in the factorization and in the back substitution are kept (i.e. A, SNR, RNR, the first 11 columns of HA, PIVOT and B). In the the second bank we keep the arrays used in the computation of residual vectors $r_i=b-Ax_i$ (i.e. A1, SN, the 12th and 13th columns of HA, X and B1). Using only one bank at each stage of the computations (either the first bank or the second one) this version will never need more storage than the storage needed for the direct solution. The price is only a modest increase in the computing time. Thus this version combines the best features of Y12MA and Y12MF. It should be mentioned that this version is machine dependent, see more details in Wasniewski et al.[1]

8.2. Runs on IBM 3033

The subroutines of package Y12M have been run on the IBM 3033 computer at the Northern Europe University Computing Center (NEUCC) at the Technical University of Denmark (DK-2800 Lyngby, Denmark). The subroutines were executed using both the FORTG and FORTH compilers. The results of these runs are given in Table 9 and Table 10. The runs were carried out in February 1981.

It should be mentioned that the subroutines could be run under the WATFIV compiler also, however the WATFIV compiler has been found less suitable for solving large algebraic systems.

1.   Wasniewski, J., Zlatev, Z. and Schaumburg, K. -
        "A Multibanking Option of an Iterative Refinement Subroutine".
        In: "Conference Proceedings and Technical Papers",
        Spring Conference of Univac Users Assotiation/Europe, Geneva, 1981.

COMPILER		FORTG				FORTH		
N	CLASS D		CLASS E		CLASS D		CLASS E	
250	2.15	(3.83)	1.58	(1.67)	1.48	(2.66)	1.08	(1.16)
300	2.66	(6.61)	2.09	(2.17)	1.87	(4.56)	1.42	(1.50)
350	2.86	(7.04)	2.61	(3.07)	1.98	(4.89)	1.80	(2.13)
400	3.26	(7.77)	3.26	(4.03)	2.26	(5.45)	2.24	(2.80)
450	3.57	(9.72)	3.88	(4.86)	2.46	(7.38)	2.66	(3.37)
500	3.90	(8.87)	4.66	(6.17)	2.70	(6.17)	3.17	(4.25)
550	4.38	(11.89)	5.23	(7.11)	2.99	(8.21)	3.58	(4.90)
600	4.65	(11.91)	5.96	(8.86)	3.16	(8.26)	4.06	(6.06)
Total	27.43	(67.64)	29.27	(37.94)	18.90	(47.58)	20.01	(26.17)

Table 9

The computing times obtained on the IBM 3033. Iterative refinement with stability factor u = 4 and drop-tolerance $T = 10^{-2}$ has been used (in brackets the results obtained by solving the system directly with stability factor u = 16 and drop-tolerance $T = 10^{-12}$ are given). Three rows with minimal numbers of non-zero elements are investigated at each stage of the Gaussian elimination. In each row of the table the sums of the computing times for the solution of six systems (with C=4(40)204 and with the same N) are given.

COMPILER	TIME	PRICE
FORTG	56.70 (105.58)	46.44 (80.52)
FORTH	38.91 (73.75)	35.04 (58.44)

Table 10

The computing times and the prices for the run of all 96 systems from Table 9 when iterative refinement (IR) and direct solution (DS) are used. The DS results are given in brackets.

8.3. Runs on CDC Cyber 173.

The 96 systems, which have been used in our experiments on UNIVAC 1100 and IBM 3033, have similarly been solved on the CDC Cyber 173 at the Regional Computing Centre at the University of Aarhus (RECAU). This computer has higher machine accuracy ($\varepsilon = O(10^{-14})$) than both UNIVAC ($\varepsilon = O(10^{-8})$) and IBM 3033 ($\varepsilon = O(10^{-7})$). This means that in the runs on CDC Cyber 173 the IR option of Y12M (subroutine Y12MFE) will use more iterations in order to achieve full machine accuracy than in the runs on UNIVAC 1100 and IBM 3033. Moreover, the double precision computations on CDC Cyber 173 are about three times more expensive than the computations in single precision (double precision computations are used in the calculation of the residual vectors in the IR option of Y12M). This shows that the part of the computational work needed to perform the iterative process is considerably increased when the CDC Cyber 173 is used. Nevertheless, the total computing time is very often smaller when iterative refinement is applied (some results are given in Table 11).

It should be mentioned here that the speed of the IR option can be slightly increased if the iterative process is terminated when a prescribed accuracy, say 10^{-5}, is achieved. Subroutine Y12MFE can in an obvious way be modified to terminate the iterations when some desired accuracy is achieved.

It is a pleasure for us to aknowledge the support recieved from the scientists at RECAU. Especially we thank dr. H.O. Andersen for his valuable help during the computations.

	DIRECT SOLUTION		ITERATIVE REFINEMENT	
N	CLASS D	CLASS E	CLASS D	CLASS E
250	20.82	8.87	14.14	9.96
300	35.82	11.69	16.55	13.22
350	39.30	16.27	17.96	17.08
400	43.97	22.02	20.72	21.46
450	54.02	26.67	22.38	26.45
500	57.53	36.43	25.20	32.48
550	64.04	38.93	27.10	37.09
600	68.46	48.08	29.88	42.98
Total	383.96	208.96	173.93	200.72

Table 11

The computing times obtained on the CDC Cyber 173. The IR option has been used with stability factor $u=4$ and drop tolerance $T=10^{-2}$. The DS option has been used with stability factor $u=16$ and drop-tolerance 10^{-12}. Three rows with minimal numbers of non-zero elements are investigated at each stage of the Gaussian elimination. In each row of the table the sums of the computing times for the solution of six systems (with $C=4(40)204$ and with the same N) are given.

9. References

1.   Bjorck, Å. -

     "Methods for Sparse Linear Least-Squares Problems".

     In: "Sparse Matrix Computations"

     (J.Bunch and D.Rose, eds.), pp.177-199.

     Academic Press, New York, 1976.

2.   Clasen, R.J. -

     "Techniques for Automatic Tolerance in Linear Programming",

     Comm. ACM 9, pp. 802-803, 1966.

3.   Cline, A.K., Moler, C.B., Stewart, G.W. and Wilkinson, J.H. -

     "An estimate for the condition number of a matrix",

     SIAM J. Numer. Anal. 16, 368-375, 1979.

4.   Dongarra, J.J., Bunch, J.R., Moler, C.B. and Stewart, G.W. -

     "LINPACK User's Guide",

     SIAM, Philadelphia, 1979.

5.   Duff, I.S. -

     "MA28 - a Set of FORTRAN Subroutines

     for Sparse Unsymmetric Matrices",

     Report No. R8730, A.E.R.E.,Harwell, England, 1977.

6.   Duff, I.S. and Reid, J.K. -

     "Some Design Features of a Sparse Matrix Code",

     ACM Trans. Math. Software 5, pp. 18-35, 1979.

7. **Forsythe, G.E., Malcolm, M.A., and Moler, C.B.** -

"Computer Methods for Mathematical Computations",

Prentice-Hall, Englewood Cliffs, N.J., 1977.

8. **Forsythe, G.E. and Moler, C.B.** -

"Computer Solution of Linear Algebraic Equations",

Prentice-Hall, Englewood Cliffs, N.J., 1967.

9. **Gustavson, F.G.** -

"Some Basic Techniques for Solving Sparse

Systems of Linear Equations".

In: "Sparse Matrices and Their Applications",

(D.J. Rose and R.A. Willoughby, eds.), pp 41-52,

Plenum Press, New York, 1972.

10. **Gustavson, F.G.** -

"Two Fast Algorithms for Sparse Matrices:

Multiplication and Permuted Transposition",

ACM Trans. Math. Software, 4, pp. 250-269, 1978.

11. **Houbak, N. and Thomsen, P.G.** -

"SPARKS - a FORTRAN Subroutine for Solution

of Large Systems of Stiff ODE's with Sparse Jacobians",

Report 79-02, Institute for Numerical Analysis,

Technical University of Denmark, Lyngby, Denmark, 1979.

12.   **Moler, C.B.** -

"Three Research Problems in Numerical Linear Algebra".

In: "Proceedings of the Symposia in Applied Mathematics"

(G.H. Golub and J. Oliger, eds.), Vol. 22, pp. 1-18,

American Mathematical Society,

Providence, Rhode Island, 1978.

13.   **NAG Library** -

Fortran Manual, Mark 7, Vol. 3, 4,

Numerical Algorithms Group,

7 Banbury Road, Oxford OX2 6NN, United Kingdom.

14.   **Reid, J.K.** -

"A Note on the Stability of Gaussian Elimination",

J. Inst. Math. Appl., 8, pp. 374-375, 1971.

15.   **Reid, J.K.** -

"Fortran Subroutines for Handling Sparse Linear Programming Bases",

Report R8269, A.E.R.E., Harwell, England, 1976.

16.   **Schaumburg, K. and Wasniewski, J.** -

"Use of a Semiexplicit

Runge-Kutta Integration Algorithm in a Spectroscopic Problem"

Computers and Chemistry 2, pp. 19-25, 1978.

17.   **Schaumburg, K., Wasniewski, J. and Zlatev, Z.** -

"Solution of Ordinary Differential Equations.

Development of a Semiexplicit Runge-Kutta Algorithm

and Application to a Spectroscopic Problem",

Computers and Chemistry, 3, pp. 57-63, 1979.

18.   Schaumburg, K., Wasniewski, J. and Zlatev, Z. -

       "The Use of Sparse Matrix Technique in the Numerical Integration of Stiff

       Systems of Linear Ordinary Differential Equations",

       Computers and Chemistry, 4, pp. 1-12, 1980.

19.   Stewart, G.W. -

       "Introduction to Matrix Computations",

       Academic Press, New York, 1973.

20.   Tewarson, R.P. -

       "Sparse Matrices",

       Academic Press, New York, 1973.

21.   Thomsen, P.G. -

       "Numerical Solution of Large Systems with Sparse Jacobians".

       In: "Working Papers for the 1979 SIGNUM Meeting

       on Numerical Ordinary Differential Equations" (R.D. Skeel, ed.),

       Computer Science Department,

       University of Illinois at Urbana - Champaign, Urbana, Illinois, 1979.

22.   Wasniewski, J., Zlatev, Z. and Schaumburg, K. -

       "A Multibanking Option of an Iterative Refinement Subroutine".

       In: "Conference Proceedings and Technical Papers",

       Spring Conference of Univac Users Assotiation/Europe, Geneva, 1981.

23.   Wilkinson, J.H. -

       "Rounding Errors in Algebraic Processes",

       Prentice-Hall, Englewood Cliffs, N.J., 1963.

24. **Wilkinson, J.H.** -

"The Algebraic Eigenvalue Problem",

Oxford University Press, London, 1965.

25. **Wilkinson, J.H and Reinsch, C.** -

"Handbook for Automatic Computation",

Vol II, Linear Algebra, pp. 50-56,

Springer, Heidelberg, 1971.

26. **Wolfe, P.** -

"Error in the Solution of Linear Programming Problems",

In: "Error in Digital Computation" (L.B.Rall, ed.), Vol 2, pp. 271-284,

Wiley, New York, 1965.

27. **Zlatev, Z.** -

"Use of Iterative Refinement in the Solution of Sparse Linear Systems",

Report 1/79, Institute of Mathematics and Statistics,

The Royal Veterinary and Agricultural University,

Copenhagen, Denmark, 1979

(to appear in SIAM J. Numer. Anal.).

28. **Zlatev, Z.** -

"On Some Pivotal Strategies in Gaussian Elimination

by Sparse Technique",

SIAM J. Numer. Anal. 17, pp. 18-30, 1980.

29. **Zlatev, Z.** -

"On Solving Some Large Linear Problems by Direct Methods",

Report 111, Department of Computer Science,

University of Aarhus, Aarhus, Denmark, 1980.

30.   **Zlatev, Z. -**

"Modified Diagonally Implicit Runge-Kutta Methods",

Report No. 112, Department of Computer Science,

University of Aarhus, Aarhus, Denmark, 1980

(to appear in SIAM Journal on Scientific and Statistical Computing).

31.   **Zlatev, Z. -**

"Comparison of Two Pivotal Strategies in Sparse Plane Rotations",

Report 122, Department of Computer Science,

University of Aarhus, Aarhus, Denmark, 1980.

(to appear in Computers and Mathematics with Applications).

32.   **Zlatev, Z., Barker, V.A. and Thomsen, P.G. -**

"SSLEST - a FORTRAN IV Subroutine for Solving

Sparse Systems of Linear Equations (USER's GUIDE)",

Report 78-01, Institute for Numerical Analysis,

Technical University of Denmark, Lyngby, Denmark, 1978.

33.   **Zlatev, Z. and Nielsen, H.B. -**

"Least - Squares Solution of Large Linear Problems".

In: "Symposium i Anvendt Statistik 1980"

(A. Hoskuldsson, K. Conradsen, B. Sloth Jensen and K.Esbensen, eds.),

pp. 17-52.

NEUCC, Technical University of Denmark,

Lyngby, Denmark, 1980.

34.   **Zlatev, Z., Schaumburg, K. and Wasniewski, J. -**

"Implementation of an Iterative Refinement Option

in a Code for Large and Sparse Systems".

Computers and Chemistry, 4, pp. 87-99, 1980.

35.   **Zlatev, Z. and Thomsen., P.G. -**

"ST - a FORTRAN IV Subroutine for the Solution

of Large Systems of Linear Algebraic Equations with Real Coefficients

by Use of Sparse Technique",

Report 76-05, Institute for Numerical Analysis,

Technical University of Denmark, Lyngby, Denmark, 1976.

36.   **Zlatev, Z. and Thomsen, P.G. -**

"An Algorithm for the Solution of Parabolic Partial Differential Equations

Based on Finite Element Discretization",

Report 77-09, Institute for Numerical Analysis,

Technical University of Denmark, Lyngby, Denmark, 1977.

37.   **Zlatev, Z. and Thomsen, P.G. -**

"Application of Backward Differentiation Methods to the Finite Element

Solution of Time Dependent Problems",

International Journal for Numerical Methods

in Engineering, 14, pp. 1051 - 1061, 1979.

38.   **Zlatev, Z., Wasniewski, J. and Schaumburg, K. -**

"Comparison of Two Algorithms for Solving Large Linear Systems".

Report No 80/9,

Regional Computing Centre at the University of Copenhagen,

Vermundsgade 5, DK-2100 Copenhagen, Denmark, 1980.

39.   **Zlatev, Z., Wasniewski, J. and Schaumburg, K. -**

"Classification of the Systems of Ordinary Differential Equations and

Practical Aspects in the Numerical Integration of Large Systems",

Computers and Chemistry, 4, pp. 13-18, 1980.

40.   Zlatev, Z., Wasniewski, J., Schaumburg, K. -

"A Testing Scheme For Subroutines Solving Large Linear Problems",

Report No 81/1,

Regional Computing Centre at the University of Copenhagen,

Vermundsgade 5, DK-2100 Copenhagen, Denmark, 1981

(to appear in Computers and Chemistry, 5, 1981).

INDEX